R(
OF GREA
AND IRELAND

ROSES OF GREAT BRITAIN AND IRELAND

B.S.B.I. HANDBOOK NO. 7

G.G. GRAHAM and A.L. PRIMAVESI

ILLUSTRATED BY
MARGARET GOLD

Distribution maps prepared by the
Biological Records Centre
Institute of Terrestrial Ecology

BOTANICAL SOCIETY OF THE BRITISH ISLES
LONDON
1993

First published in 1993 by the
Botanical Society of the British Isles
c/o The Natural History Museum
Cromwell Road
London
SW7 5BD

© 1993 Botanical Society of the British Isles

ISBN 0 901158 22 4

Design and production by
Martin Walters and Geoff Green, Cambridge

Printed in Great Britain by
Bell & Bain, Glasgow

Contents

1.	Introduction.	7
2.	Historical background.	9
3.	Problems presented by the genus: reproduction and hybridization.	13
4.	Morphology and general characters.	15
5.	Ecology and geographical distribution.	33
6.	On collecting and pressing roses.	40
7.	Classification.	44
8.	Synopsis of classification.	45
9.	List of species and hybrids noted in the text.	47
10.	Keys.	50
11.	Descriptions and figures.	57
12.	Distribution maps.	144
13.	Vice-counties of Great Britain and Ireland.	179
14.	Glossary.	183
15.	Select bibliography and references.	187
16.	Index.	203

1. Introduction

Roses are among the most familiar and ubiquitous of plants, yet their identification is notoriously difficult and their taxonomy has always been very confused. The taxonomy of the species in this Handbook is based on the findings of previous rhodologists, modified in the light of our own field studies. Hybrid descriptions are almost entirely the result of our own observations, save in the very few instances where we have seen no specimen. Here Melville (1975) has been followed.

The descriptions and keys relate specifically to British and Irish roses; we have had no experience with European material though a perusal of Scandinavian literature indicates that, apart from a greater richness in species, the situation there is much the same.

It must be emphasized that there is usually no ideal description of a rose hybrid and that recognizing a hybrid is not a simple matter of comparing specimens with our descriptions. A knowledge of local conditions and rose populations is indispensable both in recognizing hybrids and in collecting informative herbarium material. The species in any one area are usually so few that a thorough acquaintance with them will help the local recorder to recognize the great majority of hybrids or putative hybrids as competently as a national referee.

Acknowledgements

I should like to acknowledge my debt to the late Professor J.W. Heslop-Harrison who first introduced me to a study of the genus *Rosa* in Co. Durham. I am grateful also to the late Dr R. Melville who vetted my early collections and patiently explained the aberrations of this genus on several field excursions. The idea of compiling a field-guide to the roses of Britain and Ireland, on the lines of previous B.S.B.I. publications, was originally suggested to me as far back as 1980 and Dr Melville, then retired, gave his blessing to the proposal, indicating that he would be willing to give me any help or advice that I might need. As I was heavily engaged at the time with the Flora of Durham project I was not able to give my full attention to the Handbook until the Durham Flora had been published.

In the meantime I spent some time in Leicestershire helping the Rev. A.L. Primavesi to map the roses for the Flora he was editing. Tony Primavesi became so engrossed in the genus that he was co-opted as joint author of the Handbook. During our discussions I have on several occasions enjoyed the hospitality of Ratcliffe College and I express my thanks to the Rector of that establishment for this privilege. G.G.G.

Coming later into the field as a rhodologist, I too wish to acknowledge the help and advice given by the late Dr R. Melville. I am more specially indebted to Gordon Graham, who introduced me to the study of the genus, set me on the right lines, and helped me during the difficult early stages. We both enjoyed the company and benefited from the wide local knowledge of the late John H. Chandler as we sought to advance our knowledge of roses in the Stamford area and in the old county of Rutland. A.L.P.

We should both like to thank the B.S.B.I. Publications Committee and Arthur Chater in particular who gave advice and assistance at every stage of the venture and read proofs of the whole of the book. Chris Preston also saw the proofs and offered valuable comments. A.O. Chater and C. E. Jarvis of the Natural History Museum obtained copies of much of the relevant literature; we are grateful to them for this contribution. Our thanks are also due to the many who sent specimens for vetting and thus enlarged our knowledge of the genus beyond our own rather parochial experience. It was particularly useful to examine Wolley-Dod's herbarium, which had apparently lain undisturbed in the Natural History Museum as Wolley-Dod left it.

Our thanks are due to the curators of ABRN, BM, CGE, G, LIV, LTR, NMW, OXF, PR and UPS, and to Professor G.A. Swan for the loan of specimens. Professor C.A. Stace arranged some of these loans, helped with taxonomic problems and was also instrumental in crystallizing our ideas before we began work on the book. We are also grateful to the curators of BM, K, LINN and DBN for permission to study material of *Rosa* at these herbaria.

Our thanks are due also to Douglas Kent who checked the Bibliography and helped with hybrid nomenclature and to Mr and Mrs J.M. Milner who provided us with a translation of Malmgren (1986) from the Swedish.

Declan Doogue contributed much useful information from his wide knowledge of the Irish rose flora and his large herbarium proved invaluable. We are each grateful to him for his hospitality on our several visits to his country. Several correspondents provided useful mapping records, notably Mr L.J. Margetts, Mr G.H. Ballantyne, Dr G. Halliday, Miss A.P. Conolly and Mr R. Maskew.

It proved difficult to obtain material of some of the rarer alien roses, so we are extremely grateful to Lt.-Col. K.J. Grapes of the Royal National Rose Society for permission to collect specimens from the Gardens of the Rose at St Albans. The Garden Foreman, David Bartlett, assisted in the collection of the specimens from which descriptions and drawings were made.

Mr C.D. Preston, assisted by Mrs Wendy Forrest and Miss S.E. Yates, was responsible for the preparation of the maps.

Dr R.F. Smith prepared the camera-ready copy and the work was carried out using the facilities of the Durham University Computing Service.

2. Historical background

Roses have been valued from ancient times by botanist and artist alike for the beauty of their flowers, for their scent and for their wonderful variety of form. Theophrastus gave a detailed account of an old rose in 300 B.C., whilst "the red rose described by Pliny in his 'Natural History' in 79 A.D., can have been none other than *Rosa gallica* L." (Krüssmann 1981).

However the genus has been the despair of systematists since the time of Linnaeus. The known forms were already so numerous, each one differing from the next by such minor points, that it was difficult to see any definite boundaries. Linnaeus commented in particular on the difficulty of circumscribing species, and in 1820 Lindley could state that "our knowledge of European Roses has become, by the extraordinary attention they have received, so extensive that it is impossible to doubt that limits between what are called species do not exist." (Melville 1967).

Not content with such a policy of despair several rhodologists sought, in the late nineteenth century, to bring some kind of order out of chaos – notably Crépin and Christ on the continent and Baker in Britain.

Baker in his monograph (1869) is content with eleven native species and a moderate number of varieties. Yet the *London Catalogue of British Plants* (1908) enumerates 25 native species and a large number of varieties whilst Wolley-Dod in his list of British roses (1911) limits the number of species to 11, but lists about 170 varieties and forms.

Matthews (1920) acknowledges that the practice of splitting Linnaean species into innumerable forms or micro-species has, in many instances, been carried to extreme and asks what forms are to be regarded as worthy of specific rank. He thinks that morphology alone is inadequate to answer this question and says, "only by culture, combined as far as possible with cytological study, will it be possible, I think, to determine finally the genetic relationships of the numerous micro-species into which the old well known species like *R. canina* have been split." He goes on, in this most important paper, to note the increasing number of hybrids that are being recognized in such genera as *Viola, Epilobium, Mentha* and *Salix*, saying that "the presence of hybrids renders the classification of such genera extremely difficult and it is not impossible that the difficulty of classifying roses may be largely due to hybridization and segregation, complicated, it may be, by re-hybridization". He specifically notes the forms of *R. coriifolia* (*R. caesia* subsp. *caesia*) placed in the *Subcollinae* and the forms of *R. glauca* (*R. caesia* subsp. *glauca*) placed in the *Subcaninae* and suggests that they may all be of hybrid origin.

The hypothesis that certain roses had a hybrid origin had already been advanced by Christ. He thought that *Rosa hibernica,* described by Templeton in 1803, might be a hybrid between *R. pimpinellifolia* L. and *R. canina* L. agg., though this view was not generally accepted when it was first advanced in 1875. Christ also, in 1884, suggested the hybrid origin of *Rosa involuta* Sm.

During the early part of this century the genetics of roses began to be studied by Hurst, and by Blackburn & Heslop-Harrison in England, and by Tackholm in Sweden. They were followed by Erlanson, Gustafson & Hakansson, A.P.Wylie, G.D.Rowley and Klášterský.

"Heslop-Harrison & Blackburn published the results of their research in 1921 and Tackholm in [1920 &] 1922. Both discovered polyploidy in the genus and also the characteristic and almost unique meiotic behaviour of the Section *Caninae*." (Krüssmann 1981).

The findings of these geneticists are so important that a separate chapter is devoted to them in this work. Indeed Melville (1967) can say of their discoveries that "here is the long-sought key to the tremendous variability of the Caninoid roses for which Crépin was constantly seeking."

Using this 'key', however, proved to be no easy matter. It is clear that if closely related species, or species and varieties, hybridize, the hybrid offspring will be exceedingly difficult to detect. In actual practice most authors have not, as a rule, distinguished them as hybrids, but have named them as varieties under the species which they most resembled.

Although certain well-defined hybrids (notably those involving *R. pimpinellifolia*) were accepted by authors such as Wolley-Dod, who worked extensively on the genus between 1908 and 1930, there was great reluctance to forsake the plethora of varietal names already established. Indeed our foremost modern rhodologist, Dr R. Melville, showed great reluctance, for many years, to use any other classification than that put forward by Wolley-Dod in his *Revision of British Roses* (1930-31). Whilst we are prone to criticize Wolley-Dod's work, it must be said that he brought order out of chaos and gave us a working classification which served botanists for some 40 years. His species concept remains valid today and, until extensive hybridization in this genus had been accepted by professional botanists, it is difficult to see how the multiplication of varietal names could have been avoided.

It is surprising that during the period 1930-1970, whilst our understanding of other critical genera, and in particular those which are prone to hybridization, was advancing at a rapid rate, the taxonomy of the genus *Rosa* remained virtually static. The standard British Flora, Clapham, Tutin & Warburg (1962), listed simply thirteen native species, mentioned the hybrids involving *R. pimpinellifolia*, and stated that "until other forms have been cytologically studied it seems impossible to treat the group satisfactorily taxonomically."

Melville became interested in roses when working with Magnus Pyke during the 1939-45 war years on the vitamin C content of rose hips, eventually becoming the sole B.S.B.I. referee for the genus. After attending a B.S.B.I. critical plants course in 1961, Mrs I.M.Vaughan and the Rev. G.G.Graham also became interested in roses and began submitting specimens to Dr Melville for identification. The specimens were invariably named according to Wolley-Dod's *Revision of British Roses* (1930-31), though occasionally an appended note would state: "this is probably a hybrid." The attempts to match every or any rose with Wolley-Dod's sparse descriptions seemed quite subjective and always very frustrating, and it is easy to see why so few botanists were interested in this genus.

The publication of *Flora Europaea* (1968) reopened the problem so far as Britain was concerned and indicated that there was more to *Rosa* than Clapham, Tutin & Warburg allowed. On the other hand the treatment there was extremely uneven; *R. vosagiaca (R. caesia* subsp. *glauca*) was not listed as British: *R. subcanina* and *R. subcollina* were given specific status when even Wolley-Dod had recognized them as amorphous Groups (and Matthews in 1920 listed them as possible hybrids). So this account received negative criticism in British reviews but nothing positive took its place.

Eventually Melville (1975) published his mature views in an account which proved heuristic to those who had studied roses under his guidance. In this treatment Melville listed and described all the hybrids which had been recognized by himself in Britain and Ireland, and indicated that there must be many more, but that they would be very difficult to recognize. Whilst not attempting at this stage to equate all Wolley-Dod's varietal names with hybrids he did indicate many points of synonymy.

At the same time and in the same work, Stace (1975), in his introductory chapter on hybrids in general, discussed the problem of deciding on the limits of species in genera where introgression is a common phenomenon. He quoted and amplified the opinion of Webb (1951) that, where the normal population of a species shows some signs of introgression of other species, such individuals should be accepted as part of the normal pattern of variation and be named as the species. This is particularly applicable to *Rosa*, and especially to the caninoid roses, for which it is almost literally true to say that no two bushes have exactly the same combination of characters. Such an authoritative statement has helped to make the decision as to the limits of *Rosa* species considerably easier, and with some logical basis.

The years following Melville's published views have been crucial for the authors of this Handbook. It might just be possible, we thought, to go one stage further and to equate at least some of the many varietal names, not treated by Melville, with a workable number of hybrids. The process of translating the concepts of some two hundred years into modern terms has not proved easy and necessitated a great deal of field-work as well as the examination of herbarium specimens. A considerable amount of research into the literature was also necessary. Moore in Clapham, Tutin & Moore (1987) gives a summary of the situation as it was known at that date but the full results of our work were not published until Graham & Primavesi (1990). A recent summary appeared in the *New flora of the British Isles* (Stace 1991); the present work attempts to present our mature conclusions in more detail.

3. Problems presented by the genus: reproduction and hybridization

Much of the difficulty presented by the genus *Rosa*, both to the taxonomist who tries to name roses and to anyone who tries to identify them in the field, arises from two main causes, namely the readiness with which roses hybridize, and the peculiar method of reproduction which many of them display.

In the majority of cases, pollen from any one rose is capable of fertilizing any other. Some of the resulting hybrids are completely sterile. Many of them are however at least partially fertile, and are thus capable of forming further hybrids of increasing complexity in future generations. This accounts for the often bewildering combinations of characters found in roses in the field. Before this propensity to hybridize was recognized, rhodologists were apt to give specific rank to many of these hybrids, the descriptions of which could in fact often apply only to a single bush. Nowadays this proliferation of species is not accepted, but it is still difficult to decide on the limits of a species. Indeed, a little thought might well lead one to wonder why, with such a propensity for hybridization, the taxa which are now recognized as species have not long ago disappeared, leaving a single species showing a mixture of all the available characters. There has been very little significant research into this matter. However, we can propose some likely reasons for the present situation. Some of the hybrids are at least partially sterile. The longevity of individual rose bushes may also play its part, coupled with the fact that some of the species sucker freely and hence reproduce themselves vegetatively; probably, too, those which do not produce long suckers send up fresh shoots from the base. Self-pollination is possible if cross-pollination fails, and it is also possible that pollen from another bush of the same species has a competitive advantage in fertilization over that of another species. The final possibility is facultative apomixis. Opinions are divided here. Some authorities claim that roses can and do sometimes reproduce apomictically, others emphatically state that they cannot. Research is urgently needed here.

In genera where promiscuous hybridization occurs, it is considered permissible to allow for some degree of introgression when determining the limits of variation in a species (Webb (1951) and Stace (1975)). This is essential in the case of *Rosa* if we are to be able to record species at all; the alternative would be to revert to the former multitude of named species of dubious credibility. Thus we must accept that *Rosa* species are variable, and that a species in one part of the country may differ somewhat in characters from the same species elsewhere, because the gene pools available for introgression are different. After experience has been gained, straightforward first generation hybrids can usually be recognized, but complex mixtures are certain to be found, some of them of such complexity that no-one could determine their ancestry. Such plants should be ignored for recording purposes – there are always plenty of roses which can be determined and recorded.

The other problem which complicates the study of *Rosa* is almost unique to the genus. *Rosa arvensis, R. pimpinellifolia* and all of the alien species described in this book reproduce in the normal way, half of the genetic material being supplied by each parent. For the rest of the British species, all in fact belonging to the Section *Caninae*, the majority of the inheritable characters, four fifths in most cases, are determined by the seed parent. Also when the seed parent belongs to this Section, three fifths of these characters are transmitted unchanged to successive generations. This may be a further factor in maintaining the stability of species. As a result of this peculiar method of reproduction, reciprocal hybrids tend to be matriclinal in characters, resembling the seed parent more than the pollen parent; this difference in characters between the two possible hybrids is especially marked when one of the parents belongs to the Section *Caninae* and the other to another Section of the genus. A more technical explanation of this phenomenon follows:

The species of Section *Caninae* are all unbalanced polyploids. As can be seen in the synopsis of classification most of them are pentaploids, so to avoid circumlocution this level of polyploidy will be used as an example.

The basic chromosome number for the genus *Rosa* is 7. In a pentaploid rose there are 7 bivalents (14 paired chromosomes) and 21 univalents (three sets each of 7 unpaired chromosomes). Only the 14 paired chromosomes take part in the process of meiosis. In the formation of the pollen grain, the 21 unpaired chromosomes are lost; in the formation of the embryo-sac they are retained. Thus the pollen parent provides 7 chromosomes, the seed parent provides 28, and on fertilization the pentaploid number 35 is restored.

It can be seen from the above why hybrids between two species, both of which belong to Section *Caninae* are matriclinal in characters, and why the seed parents of any number of future crosses will carry many of the characters of the original seed parent unchanged. When only one of the parents is an unbalanced polyploid the situation is a little more complicated. Thus, for example, *R. canina* × *R. arvensis,* with *R. canina* as the seed parent, is pentaploid, with four fifths of the genetic material provided by *R. canina*; the reciprocal hybrid, *R. arvensis* × *R. canina*, is diploid, with half the genetic material provided by each parent. In a hybrid between *R. pimpinellifolia* and a pentaploid species, the hybrid will be triploid if *R. pimpinellifolia* is the seed parent, and hexaploid if it is the pollen parent.

For anyone wishing to follow this matter further, the appropriate references are given in Chapter 2 and in the bibliography.

4. Morphology and general characters

For many people the word 'rose' conjures up a picture of a large many-petalled flower, or of garden roses with growth forms artificially induced or selected to suit various situations in gardens. These are often widely different from our native British roses, so it must be understood that the following descriptions are of wild roses only, and of such alien species as may be found growing in the wild and which are described in this Handbook.

The genus *Rosa* is acknowledged to be difficult, both taxonomically and morphologically, and most people when they first try to get to grips with it can find themselves bewildered or even discouraged. Of the native British species, only *R. arvensis* and perhaps *R. pimpinellifolia* show clear-cut and invariable characters, and hence are easy for all to recognize. Any specimen with many of the characters of one of these two species, but differing significantly from the normal, can be assumed to be a hybrid. All the other species are variable, some of them extremely so. Indeed, it is not far from the truth to say that apart from the two species just mentioned, no two rose bushes have exactly the same characters. Some of the reasons for this variability are discussed in the previous chapter. When attempting to identify a specimen one is often faced, not so much with a selection of clear-cut characters, as with a combination of characters which it needs judgement and some experience to interpret. The situation is further complicated by the virtual certainty of encountering hybrids, including plants in which there is slight introgression of another species – for instance it is not unusual to find otherwise good specimens of *R. canina* or of *R. obtusifolia* with a few stipitate glands on the pedicels. Beginners in the study of *Rosa* should not be discouraged by these difficulties. These will cease to be a problem when they have become familiar with the species of their own regions. Ideally the best way of doing this is to have a few outings with a knowledgeable person. Failing this, the B.S.B.I. referees are always ready to help with determinations and advice.

Habit

British roses are all shrubs with comparatively slender stems, never attaining the proportions of a tree, as may occur, for example, in species of *Crataegus* or *Prunus*. The habit of the bush, whether arching, trailing or erect, is a useful diagnostic character. However it should be emphasized that these habits may be considerably modified or obscured either by environmental conditions or by interference from man or animals. A climbing and arching species in hedgerow or woodland is easily recognizable as such if it has full opportunity to climb, but in dense shade or if partially smothered by other plants it may be stunted so that the habit is not so apparent. Species with a normally upright habit in similar situations may not easily be recognized as such. In heavily-trimmed hedgerows the habit of the bush is often hopelessly modified. In open situations species with erect habit show this habit at its best, but as stated below, in such situations climbing species may not develop the dome-shaped appearance described for some years, and may remain upright with little sign of arching stems. The habit of the bush is obviously not apparent in herbarium specimens or in small pieces sent to a referee for determination.

Figure 1. Habit. (a) Climbing and arching. (b) Arching, unsupported. (c) Trailing. (d) Trailing, supported.

It is therefore good practice to annotate the label of a herbarium specimen with a description of the habit and it is extremely useful for a referee if this information is sent with the specimen. There are two main growth forms:

1. *Climbing and arching* (Fig. 1). Shrubs with young stems which are too weak to support themselves in an upright position, and which bend under gravity into an arching shape. These are the climbing species, often found in hedgerows or in other situations where there is support from other plants. In such situations there are usually several fairly thick stems

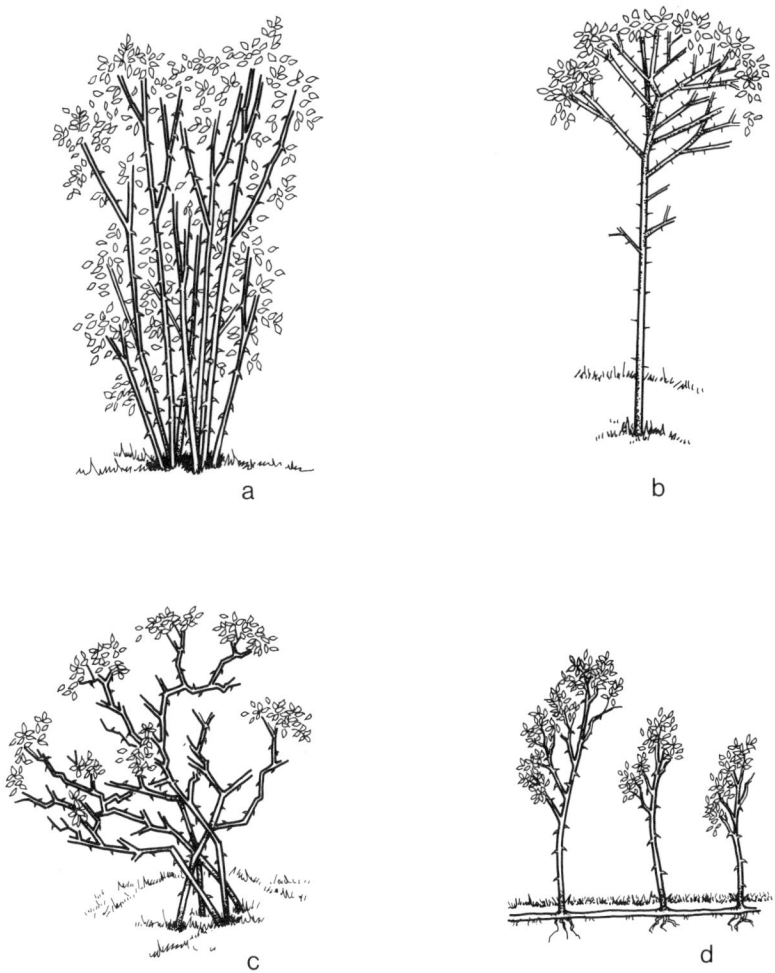

Figure 2. Habit. (a), (b) Erect, with straight stems. (c) Erect, flexuous. (d) Sucker shoots.

arising from ground level, erect or at a slight angle, and then branching and spreading haphazardly according as the hedgerow shrubs give them support. The free ends of the young stems assume the arching shape (Fig. 1a). *R. canina* is a common example.

Such species, if growing in open situations, will at first form a small upright bush, and if there is much interference either from man or from animals which inhibits the growth of long arching stems, will retain this upright habit. If left undisturbed in open situations these species will ultimately form a wide dome-shaped bush with stems arching out from the centre (Fig. 1b). The extreme case of this weak-stemmed habit is *R. arvensis*. When growing unsupported, as this species often does, the long weak stems, though not completely decumbent, are trailing rather than arching so that the bush is a wide and low sprawling mass (Fig. 1c). When supported by other plants the ends of the long weak stems often hang vertically downwards (Fig. 1d).

2. *Erect* (Fig. 2). Shrubs with young stems strong enough to be little affected by gravity, so that the habit of the bush remains throughout life erect and compact. These bushes usually arise from a single stem at ground level. If this stem almost immediately begins to branch, the bush takes on the form shown in Fig. 2a. If the young stems of the species do bend slightly under gravity, a broader bush is formed with the appearance of Fig. 2b. In this case the young stems are flexuous rather than arching. In other species, notably *R. mollis* and *R. agrestis,* the stems do not branch immediately from ground level, giving the appearance of a small tree with a prickly trunk (Fig. 2c.) In old bushes of *R. mollis* the 'trunk' may attain a diameter of over 5 cm. Finally there are species (notably *R. pimpinellifolia, R. mollis* and *R. rugosa)* which spread from suckers just below ground level (Fig. 2d), and which sometimes form dense or even impenetrable thickets.

Prickles.

It is characteristic of *Rosa* for the stems to be armed with prickles, though occasionally one comes across a bush which is nearly or completely unarmed. These prickles are epidermal outgrowths, not thorns like those of *Crataegus* or *Prunus* which are modified branch stems. The shapes and size of the prickles are often very useful diagnostic characters, not only for determining species but also for deciding on one of the parents of a hybrid. Four basic shapes are described here, but in practice they may well be found to merge into one another in an almost continuous range of gradations. The shape and size of prickles will be found to vary in different parts of the same bush. It is important when attempting to identify a specimen to examine the prickles on the more mature parts of the stem. The prickles near the extremities of the young shoots are not always characteristic of the species, presumably because they are still immature and have not yet acquired their final shape. The prickles on the very oldest parts of the stem may also be untypical. In hybrids, there may be prickles characteristic of each parent situated on different parts of the bush. In this as in other matters there is no substitute

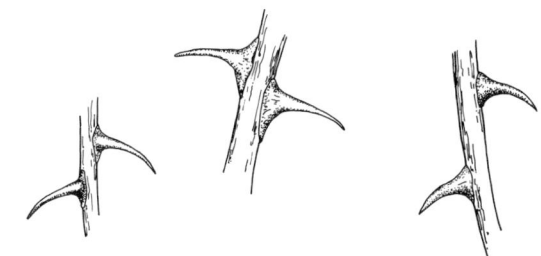

Figure 3. Prickles. Slender and curved.

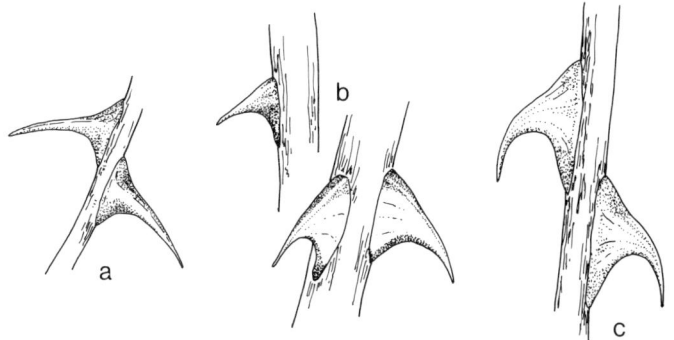

Figure 4. Prickles. (a) Transitional. (b), (c) Stout and hooked.

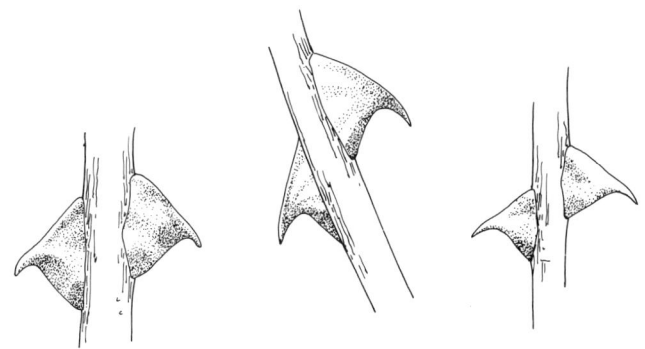

Figure 5. Prickles. Deltate.

for experience. The descriptions become more significant when one begins to become familiar with the roses of one's own region.

1. *Slender and curved* (Fig. 3). These arise from a short and narrow elliptical base. The convex and concave sides which form the curve are so disposed that the prickle tapers gradually towards its point, but the cross section is more or less elliptical throughout. There is no clear demarcation between this type and the next, intermediate forms being often found.
2. *Stout and hooked* (Fig. 4). This type arises from a large and broadly elliptical base. The degree of hooking depends, of course, on the curvature of the two edges, and especially on that of the lower edge. For example, if the lower edge forms an almost complete semicircle, as in *R. obtusifolia* and *R. rubiginosa,* the point of the prickle is directed vertically downwards forming a very strong hook (Fig. 4c). Where the curvature of the lower side is only slight, the prickle may be approaching the shape of type 3 following.
3. *Deltate* (Fig. 5). These prickles are broad based, with the upper and lower curvatures so slight as to make the prickle almost triangular in side view, resembling the Greek letter capital delta. There is only one species, *R.*

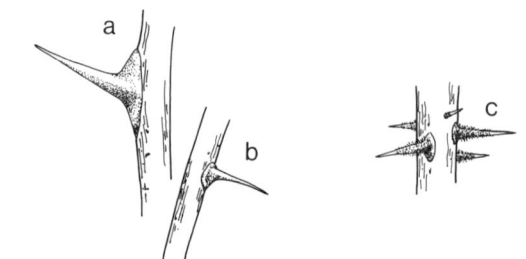

Figure 6. Prickles. Straight. (a), (b) *R. mollis.* (c) *R. rugosa.*

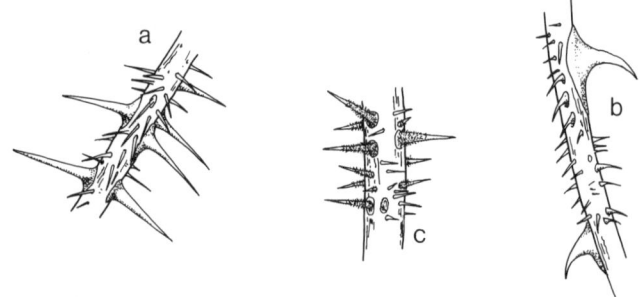

Figure 7. Mixed prickles and acicles. (a) *R. pimpinellifolia.* (b) *R. rubiginosa.* (c) *R. rugosa.*

stylosa, in which this character is really obvious, and tendencies towards this shape are often indicative of hybrids of this species.

4. *Straight* (Fig. 6). This type is a distinct one. However it is particularly important in this case to examine the prickles on the mature stems. The soft prickles on the younger stems may become bent, especially in transit when a specimen is taken. A herbarium specimen of this type should most certainly include a portion of mature stem. These straight prickles are slender in outline like sharp-pointed spines, and are nearly circular or slightly elliptical in section. The edges are straight until just above the base, where the curvature increases. The base of the prickle is more or less circular or slightly elliptical, modified sometimes by being prolonged into a long narrow pear-shape as in Fig. 6b. Only two species described in this book have straight prickles, namely *R. rugosa* and *R. mollis.* The prickles of *R. rugosa* are distinctive in that they are hairy towards the base.

5. *Acicles.* In a few species such as *R. pimpinellifolia* the larger prickles may be interspersed with much smaller ones called acicles (Fig. 7). These are often straighter than the large prickles, and may come in all sizes down to what appear to be stiff bristles. However unlike a true bristle (or stiff hair), acicles always taper to a point from a broader base. Occasionally acicles are gland-tipped.

6. *Pricklets.* Mature prickles on the stems are usually all more or less of the same size. However, many species have smaller prickles, or pricklets, on the under sides of petiole and rachis. These differ from acicles in that they closely resemble the main prickles and are all more or less uniform in size.

It will be obvious from what is said above that the difference in curvature of upper and lower sides of prickles results in many different overall shapes, some of them very characteristic of particular species. It is virtually impossible to describe these curvatures and their combinations in words which would not be liable to misunderstanding. In the accounts of species and hybrids, therefore, reference should be made to this chapter or to the appropriate species drawing.

Leaves.

The leaves of *Rosa* provide many important diagnostic characters. There are other characters which may be of greater diagnostic importance, but a quick look at the leaves is often an effective way of narrowing down the number of species at the beginning of the investigation.

The leaves of British *Rosa* species are imparipinnate, that is, the leaf is compound with 2-5 pairs of opposite leaflets and a single terminal leaflet (Fig. 8). At the base of the petiole there is a pair of narrow stipules adnate to the petiole over most of their length, but with free portions called auricles at their tips. The petiole and rachis are often, but not always, armed on their under sides with pricklets.

The leaves provide the following useful diagnostic characters:

1. *Size.* The size of leaves and leaflets may vary considerably in different

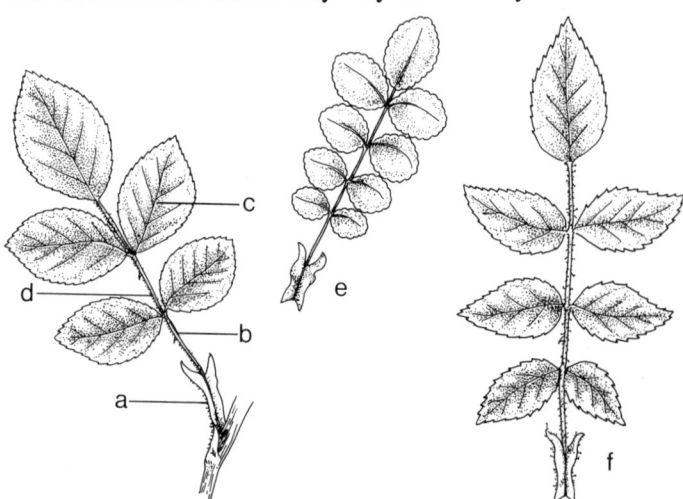

Figure 8. Leaves. (a) Stipules. (b) Petiole. (c) Leaflet. (d) Rachis. (e) Leaflets contiguous. (f) Leaflets widely spaced on rachis; lowest pair reflexed.

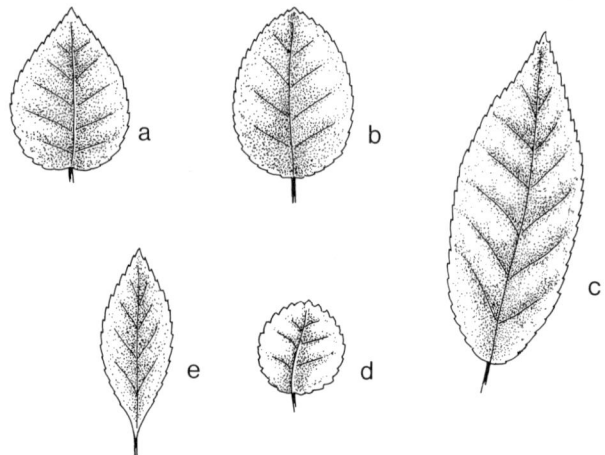

Figure 9. Leaflet shape. (a) Ovate, acute. (b) Ovate, obtuse. (c) Ovate-lanceolate. (d) Orbicular. (e) Cuneate at base.

parts of a single bush, or may be influenced by the type of habitat in which the bush is growing. However it is possible to describe these in general for a particular species as large or small without misleading.

2. *Shape.* (Fig. 8). The overall shape of a leaf is a somewhat elusive quality which is difficult to describe, but which, when experience of local populations has been gained, often provides valuable clues. It depends largely on the lengths of petiole and rachis, the relative size of the leaflets and their spacing and disposition on the rachis, together with the other characters of the leaflets described here. For example, the leaves of *R. pimpinellifolia,* with up to 5 pairs of small rounded leaflets set close together on the rachis are unmistakable; the subtly neat appearance of the leaves of *R. obtusifolia,* once seen, is never forgotten; the widely spaced uniserrate leaflets of *R. stylosa,* with the lowest pair somewhat reflexed, are unique to that species. There are also other subtle characters such as the tendency for the leaflets to fold in *R. caesia* subsp. *glauca,* but it is best at first not to rely too much upon these impressions until one has acquired some familiarity with the taxa.

The shape of the leaflets is more easily described, and is often a useful diagnostic character (Fig. 9). Basically they are all ovate, terminating in a point which may be acute or obtuse, but this basic shape may be modified from ovate-lanceolate to suborbicular in outline. The bases of the leaflets are usually rounded, but occasionally, as in *R. agrestis* and its hybrids, they may be cuneate.

3. *Indumentum.* Some species have entirely glabrous leaves. In others, there may be hairs on petiole and rachis; the leaflets show a range of hairiness from densely tomentose on both surfaces to entirely glabrous or with a few sparse hairs confined to the midrib and veins beneath. It is best to examine the lower surfaces of the leaflets first, and for this and the following two characters a good hand lens, preferably of × 10 magnification, is almost essential. It should be noted that a specimen with a few sparse hairs on the leaflets may provide a pitfall, especially when the presence of these hairs does not seem to conform with other diagnostic characters. This is often a sign of introgression. If the specimen appears to conform with a named taxon in all other respects, this slight anomaly can be assumed to fall within a permissible degree of variation. Otherwise it is best to ignore such specimens for recording purposes, or to suspect hybridity.

4. *Serration.* The margins of the leaflets in all species are all sharply toothed. In most species they are serrate, that is, having sharp teeth with more or less straight sides. In some species they are crenate-serrate, the teeth having rounded sides but ending in a sharp point. It is important to note, however, that the teeth near the base of the leaflet are often not

characteristic of the leaflet as a whole. For example, the basal region of the leaflets of *R. rugosa* is truly crenate, with no sharp points, the rest of the leaflet being crenate-serrate.

For descriptive purposes the serration of the leaflets can be divided into three types:

(a) *Uniserrate* (Fig. 10). A uniserrate leaflet has sharp teeth, all more or less the same size (though they may decrease in size towards the base of the leaflet). In some cases, notably *R. canina* Group *Transitoriae* and sometimes *R. caesia*, the leaflet is *irregularly uniserrate*, with a

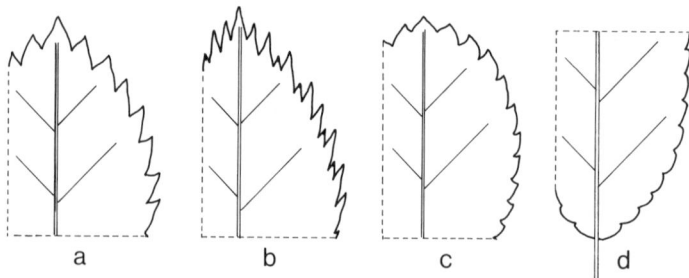

Figure 10. Uniserrate. (a) Regularly uniserrate. (b) Irregularly uniserrate. (c) Crenate-serrate. (d) Crenate.

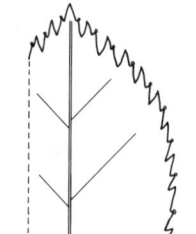

Figure 11. Biserrate.

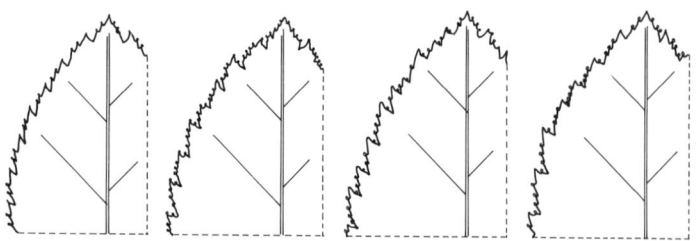

Figure 12. Multiserrate.

few smaller teeth spaced at irregular intervals between the larger ones. These smaller teeth are often tipped with dark-coloured rudiments of glands. This should not be confused with a truly biserrate leaflet (cf. Fig. 11 following), in which the smaller teeth arise on the sides of the larger ones and not between them on the leaf margin.

(b) *Biserrate* (Fig. 11). A biserrate leaflet has large teeth, from the lower side of each of which a single smaller tooth arises. These secondary teeth are frequently gland-tipped.

(c) *Multiserrate* (Fig. 12). A multiserrate leaf has large teeth, on the lower side of which, and sometimes also on the upper side, two or more smaller teeth arise, these secondary teeth being usually gland-tipped. This type is not as well defined as the previous ones, because in some cases the secondary teeth are little more than rounded protuberances tipped with glands, and the larger primary teeth are sometimes not noticeably sharp-pointed.

Glands.

The leaves of some species are entirely eglandular. In others, the stipules may be fringed with stipitate glands, and there may be similar glands on petiole and rachis. The margins of the leaflets may have small sessile or shortly stalked glands; the occurrence of these is discussed above in the description of 'biserrate' and 'multiserrate' leaflets. There may also be glands on midrib and veins of the leaflets, or scattered over the lower surface. Care should be exercised in examining species with densely tomentose leaves as the glands may be hidden by the tomentum and hence difficult to see. Glands similar to those associated with the leaves are found also on other parts of the plant. The size, shape, colour and odour of these glands can provide important diagnostic characters. They can be divided into three types:

1. The first type can only loosely be defined as glands. They appear rather to be warty protuberances or outgrowths of tissue, differing little in cell structure from the cells of the organ whence they arise. When fully formed they are usually red or reddish-brown in colour, stalked, and with a small globose head some 40μm in diameter. However they vary considerably in shape and in length of stalk, and, particularly when they are associated with leaf serrations, may be reduced in the extreme case to a few coloured cells at the tip of a serration.

   Stalked, or stipitate glands of the first type are best seen on the margins of stipules or bracts. They also occur diagnostically on the pedicels of certain species, notably *R. arvensis* and *R. stylosa*.

2. The second type of gland is found associated exclusively with the three species *R. tomentosa, R. sherardii* and *R. mollis* and their hybrids. In this type there is less variation in form – a small globose head 40-50μm

in diameter on a short or very short stalk being the norm. Their colour varies from red to orange-yellow. When fresh, such glands are faintly translucent, their cells containing, it seems, aromatic compounds having a resinous odour which is released when the glands are crushed. This resinous odour decreases in intensity from *R. mollis* through *R. sherardii* to *R. tomentosa,* being very faint in the last case. In the herbarium, glands of this second type lose their translucence and dry out leaving behind a few specks of grey cellular tissue. When present on stipules, sepals and pedicels such glands can be seen to possess stalks. When subfoliar in position they usually have extremely short stalks and may be hidden among the leaf tomentum.

3. The third type of gland is associated exclusively with Sweet-briar species. Such glands are in effect a larger version of those of the previous section, their globose heads being 100-120μm in diameter, their stalks longer and thicker. The contrast between the three types of gland is best seen when subfoliar examples of each can be compared. Sweet-briar glands are usually of a golden or brownish tinge and, when fresh, viscid or very translucent. They give off a strong fruity or ripe-apple odour. On a warm, still day the scent from bushes of *R. rubiginosa* can often be detected at a distance. In the case of *R. micrantha* it is usually necessary to crush the glands in order to release the scent. With *R. agrestis* the scent is still harder to detect. It is worth emphasizing, however, that some people, even if they have an otherwise excellent sense of smell, cannot detect the fruity odour of the Sweet-briar glands. The greyish-cream specks left behind when Sweet-briar glands dry out are more noticeable than in parallel cases when glands of the second type are noted on dried specimens.

Flowers. (Fig. 13)

As a general rule, it is difficult if not virtually impossible to make a firm determination of a specimen of *Rosa* at flowering time. It is true that some species have distinctive flowers; for example the flowers of *R. arvensis* and *R. rugosa* are unmistakable once seen. But the flowers of the majority of the taxa described in this book are very similar in general appearance, and the various parts which provide diagnostic characters are of little use, until the fruit develops. Therefore only a brief description of the structure and arrangement of the flower is necessary here.

The flowers of British rose species have five sepals, usually green, but often purplish in *R. arvensis,* which act as a protective sheath when the flower is in bud, and five petals, ranging in colour from white through shades of pink to deep red. (There are some foreign species, occasionally grown in gardens, which have four sepals and petals). The stamens are very numerous, standing out in several whorls within the ring formed by the base of the petals. Beneath the base of the sepals lies the hypanthium, or concave receptacle, which will develop into the hip after anthesis. Within this concave receptacle is the ovary, consisting of a number of free carpels, each with its separate style. The styles are either free throughout their length or fused in their upper part into a column. In either case, the ends of the styles are aggregated into a head of stigmas. The styles emerge through an orifice at the top of the hypanthium. Except in Section *Synstylae* and in *R. stylosa* where the styles protrude in a column, the head of stigmas at flowering time is often hidden by the mass of stamens.

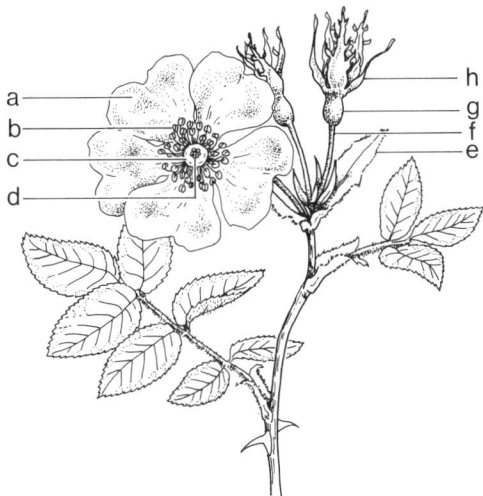

Figure 13. Flower. (a) Petal. (b) Stamens. (c) Disc. (d) Stigmas. (e) Bract. (f) Pedicel. (g) Hypanthium. (h) Sepals.

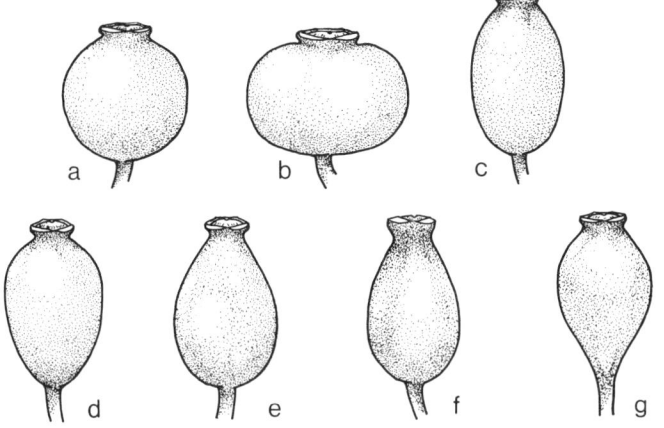

Figure 14. Hip shapes. (a) Globose. (b) Depressed-globose. (c) Ellipsoid. (d) Obovoid. (e) Ovoid. (f) Urceolate.

27

The flowers are usually in a corymbose inflorescence, though in a few cases they are solitary.

Hips. (Fig. 14)

After anthesis the concave receptacle develops into a pseudocarp or false fruit called a hip. This is a hollow structure with fleshy walls enclosing the true single seeded fruits or achenes. The persistent styles from the achenes emerge through an orifice at the apex of the hip, and are crowned with a head of stigmas; it should be noted however that often very late in the season the stigmas deteriorate so as to be unrecognizable. Also at the apex of the hip and outside the rings of withered stamens are the whorls of sepals, usually deciduous before the hip is fully ripe, but in a few cases persisting until it rots. The hip is carried on a pedicel which in most cases is subtended by a green bract.

The hips and their associated structures are extremely important in determinations. It is not necessary to wait until they are fully ripe. They acquire their final size and shape whilst they are still green, and at this stage, or in the early stages of ripening, there will still be sepals attached even in those species where the sepals fall early.

In some species, such as *R. arvensis,* the size and shape of the hips are fairly constant characters; indeed in this particular species abnormally large or unusually shaped hips are usually a sign of hybridity. In other species these characters are somewhat variable, and in *R. canina* in particular they are extremely so. Therefore unless especial emphasis is given to them in a species account it should be assumed that these characters are of minor importance compared with the others described.

The shapes of the hips can often be given definitive names such as globose, ovoid, obovoid, urceolate, etc., and these are illustrated in Fig. 14.

Occasionally one comes across a bush where the hips are abortive, or so misshapen as to be obviously partially sterile. Provided that there is no evidence of disease in the bush this is a sure sign of hybridity. Hybrids with *R. arvensis* as the female parent are particularly likely to show this phenomenon.

Pedicels.

In spite of the fact that the lengths of pedicels may vary considerably on a single bush, the comparative average length is usually a useful character in distinguishing between species. A little thought will show that variation in the length of pedicels is inevitable in a corymbose inflorescence. In order that the flowers should not hopelessly overlap, the central pedicels of a many-flowered corymb must be shorter than average, and the outer ones longer.

The presence or absence of glands on pedicels, hips or sepals is frequently of importance in diagnosis. The glands on pedicels and hips are usually stipitate;

they are occasionally accompanied by acicles which may or may not be gland-tipped. Occasionally one comes across a bush which agrees perfectly with the characters of a particular species except that there are a few sparsely scattered glands on the pedicels, sometimes only one or two glands on a few pedicels, the rest being smooth. This is a sign of introgression of another species, and if the glands are very few in number the specimen can be considered to come within the permissible range of variation for the species.

Bracts. (Fig. 13e)

A bract is a leaf in the axil of which a flower bud develops, and is frequently modified so as not to resemble the other leaves.

In British species of *Rosa* the bracts are ovate or ovate-lanceolate and acute, usually with entire margins but sometimes fringed with stipitate glands. They are usually inconspicuous, and because they seldom have much significance as diagnostic features they will not be found mentioned in many of the species accounts. In *R. pimpinellifolia* bracts are entirely absent (or perhaps the flower buds arise in the axils of normal foliage leaves). In *R. caesia* and some of its hybrids the bracts provide a useful diagnostic character because they are large, broad and conspicuous, and tend to conceal the short pedicels.

Sepals. (Fig. 15)

1. *Shape*. Sepals may be entire, with smooth edges, or they may be pinnately lobed. When entire, they range in shape from short and broadly triangular to long and linear. A curious arrangement is found in those species which have pinnately lobed sepals. (Fig. 13). Only two of the five sepals are completely lobed on both sides. Between these two on one side is a sepal which is entire, without lobes, and between them on the other side

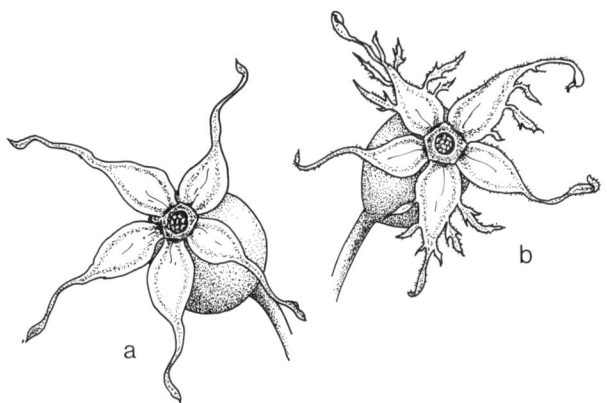

Figure 15. Sepals. (a) Entire. (b) Lobed or pinnate.

there is one unlobed sepal, and another which is lobed on one side only. This is not immediately obvious without close inspection, because the fully lobed sepals are the most conspicuous and catch the eye first. To save verbiage, in the descriptions of species with lobed sepals, only the fully-lobed sepals are described. This peculiar character, possessed by the sepals of all the caninoid roses, forms the answer to a mediaeval Latin riddle. We give two versions of the riddle and the translation as noted in Stearn (1966 p.494):

The riddle of the five brothers

Quinque sunt fratres, duo sunt barbati,
Sine barba sunt duo nati,
Unus ex his quinque,
Non habet barbam utrinque.

or

Quinque sumus fratres, unus barbatus et alter
Imberbesque duo, sum semiberbis ego.

"Five brethren of one birth are we,
All in a little family,
Two have beards and two have none,
And only half a beard has one."

Transl. E.B.Cowell.

2. *Persistence.* In some cases the sepals fall very early whilst the hips are still green; in others they persist at least until the hips begin to redden, often longer. In both these cases there is a constriction at the base of the sepals which marks the region of abscission. This constriction is not always easy to see. A better guide is that if the sepals are destined to fall, early or later, they are thin and papery right to the base, and in the later stages brown and dead-looking.

Of the British species only *R. mollis* has permanently persistent sepals, and in this case the bases of the sepals are thick and fleshy, and when the hip is ripe turn red as if they were prolongations of the hip itself.

3. *Disposition* (Fig. 16). After the fall of the petals and as the hip ripens the sepals assume an attitude in relation to the hip which is characteristic of each species, and can provide a useful character for determination. This may range from very strongly reflexed, so that the sepals are pressed against the hip, to stiffly and more or less vertically erect. Sepals which fall very early are usually strongly reflexed. In the case of the longer-persistent sepals, it should be noted that in early stages of maturity of the

hip sepals may rise somewhat before assuming their ultimate attitude, and of course shortly before they fall when the attachment is weak they may be disposed anyhow. On any one bush there are usually hips in various stages of maturity. It can be assumed that the ultimate attitude will have been assumed when the hips have thoroughly reddened.

Styles and stigmas.

Each achene has its persistent style terminating in a knob-like stigma. All these styles come together as, usually, a loosely aggregated column which emerges through the orifice of the hip. In a few cases the styles are fused to form a true single column; in *R. stylosa* they are connivent at first so as to appear to form a column, but soon separate. In most cases the styles terminate so that the stigmas form a head just above the surface of the disc, or sunk into it if the disc is strongly concave. In others they may be conspicuously exserted. These dispositions of styles and stigmas can be of considerable importance in diagnosis.

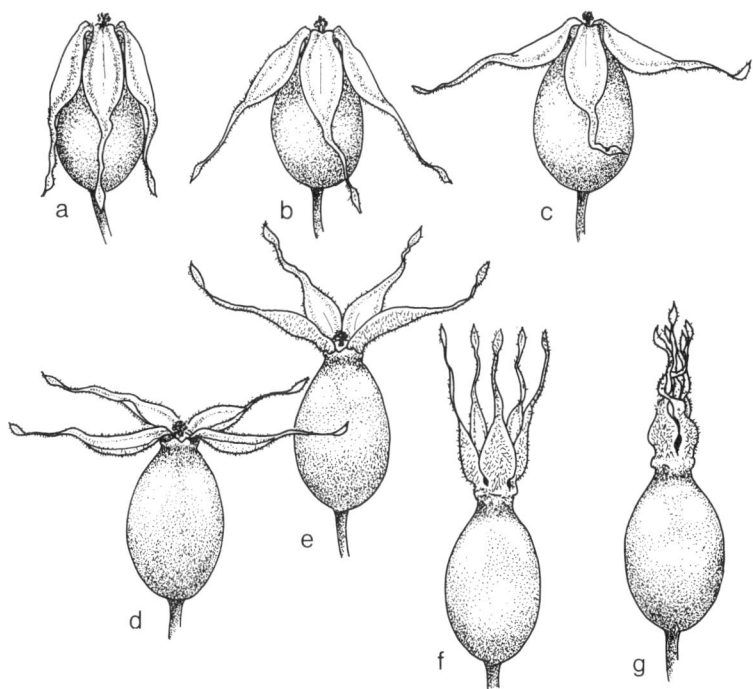

Figure 16. Disposition of sepals. (a) Appressed to hip. (b) Reflexed. (c) Reflexed-spreading. (d) Patent or spreading. (e) Spreading-erect. (f), (g) Erect.

The styles may be glabrous or hairy, and this is often a useful diagnostic character. In many species, of course, only the heads of stigmas are externally visible, and the styles cannot be seen without dissecting the hip. Fortunately this is not necessary, because the hairs near the ends of the styles, if present, can always be seen interspersed with the stigmas, which are not themselves hairy. Thus when in the description of a taxon it is stated that the styles are hispid, what one actually observes is a head of stigmas interspersed with short stiff hairs. If the description loosely refers to a villous head of stigmas, this is to save a longer description – in fact the woolly mass of hairs arises from the ends of the styles, not from the stigmas themselves.

Disc and orifice.

The disc is the region at the apex of the hip delimited by the innermost ring of stamens. The shape of the disc, whether concave, flat, convex or conical, is often a useful character. The centre of the disc is perforated by the orifice through which the styles emerge. In closely related species the ratio of diameter of disc and diameter of orifice is often an important distinguishing character. For convenience in observing the sizes of the orifice, the styles should be pulled out and the disc and orifice examined with a hand lens. The orifice may have a curved lip so it must be emphasised that it is the width of the inner cylinder, through which the styles protrude, that is being estimated against the total width of the disc. In other words estimate or measure the orifice at its narrowest part. Once experience has been gained actual measurement is rarely necessary, but it may help as a clinching factor in the determination of several hybrids, e.g., *R. sherardii* × *R. tomentosa*.

5. Ecology and geographical distribution

In Britain and Ireland at the present day there is a predominance of man-made habitats, and there are very few, if any, which are not modified in one way or another by human activities. Hence we can only speculate as to what sorts of natural habitats would have favoured the British species of *Rosa*, and from which they presumably originated, before human activities grossly affected the environment.

In general, the British *Rosa* species can be placed in two categories, namely the species with young stems too weak to support themselves, and which therefore grow best in habitats where there are other plants which they can use to provide support, and those with stems strong enough to grow upright without requiring support from other plants. This distinction must not be carried too far, but it may provide a basis for speculation as to what are the truly natural habitats for our *Rosa* species.

The most obvious natural habitat for the climbing species of *Rosa* is woodland. However, few, if any, of our roses can tolerate dense shade, and it is mainly on the margins of woodland or in the more open spaces such as the edges of rides that we find most of the roses in this habitat. *R. arvensis*, a typical woodland species, seems to be tolerant of a fair amount of shade. Of all our species this has the weakest stems, and so one would expect that it would be in the greatest need of the support of other plants in order to reach the light. It certainly does climb, sometimes to considerable heights, with the ends of its long weak stems hanging vertically downwards. This is usually in places where there is fairly dense shade. Its more usual habit is to form low sprawling masses without support in more open parts of the wood, perhaps taking advantage of its tolerance for partial shade to colonize these particular areas. Species of *Rubus* colonize open spaces in woodland in the same way.

The many miles of hedgerows, which certainly in their origin form an artificial habitat, provide an ideal extension of the natural environment described above. The climbing roses will here find the support which they require, and perhaps the partial shade at the base of the hedgerow may provide stimulus for elongation of the stems. Unfortunately the grubbing up of hedgerows which has been so prevalent in recent years has severely reduced this habitat, and so the frequency of *Rosa* species. Another adverse factor is the brutal fashion in which hedges are now trimmed. The traditional and civilized method of laying hedges and then letting them grow up for a few years before trimming did not prevent the climbing roses from adopting their normal habit of growth. When, as now, the hedges are drastically slashed with flails all the young leading shoots of the roses are destroyed, and thus they are often prevented from flowering and fruiting.

Within its geographical range, *R. arvensis* is common in hedgerows in close proximity to woodland, and decreases in frequency as the distance of the hedgerows from woodland increases. It may appear that too much prominence has been given to this species in these notes, but its autecology seems to be of particular interest, and it may throw light upon the ecology of *Rosa* as a whole. For example, the distribution map of *R. arvensis* in the *Flora of Leicestershire* (Primavesi & Evans 1988) shows that it is absent from the lower valleys of the two principal rivers in the county, the courses of the rivers showing clearly on the map as white spaces among the dots. An almost identical distribution pattern is shown by *Tamus communis*. No really satisfactory reason for this phenomenon has been proposed; the most likely reason is the absence of extensive woodland in these valleys.

Species of *Rosa* are absent or of rare occurrence in very wet ground such as marshes, fens or bogs. They also, with the possible exceptions of *R. mollis* and *R. sherardii,* seem to avoid very acid soils. Thus roses of any sort are comparatively rare in moorland in the north, regardless of whether the soil is wet or dry. Similarly there are few roses to be found on the acid heaths in the south. This is particularly noticeable in parts of Surrey where chalk gives way to greensand. There is a dramatic change in the general vegetation within a space of 100 metres or so. It is noticeable that species of *Rubus* appear to have no aversion to such soils, and that areas of interest to a student of brambles are often unprofitable to the rose enthusiast. There are hedgerows on more or less neutral soils where *Rubus* and *Rosa* grow happily together, but in general there does appear to be a difference in soil preferences. The diversity or otherwise of species occurring in hedgerows is perhaps not primarily related to soil types, and much has been written on this subject. It would be interesting to know why, for example, hedges formed by suckering elms, still fairly common in the Midlands though the trees have succumbed to disease, are often entirely devoid of roses.

In open habitats where the soil is suitable one would expect the erect-growing species which do not require support to be most at home, and this is indeed the case. In nature these open spaces on comparatively dry soil are rarely permanent, but merely phases in a succession leading to a climax vegetation, usually woodland. In this context the roses might be considered as opportunists or colonists taking advantage of a particular temporary set of conditions which suits them. In other words, they are among the makers of this particular stage in the succession, and as such they are preparing the ground for other species to carry the succession further, thus dooming themselves to ultimate extinction in this particular locality. Usually the persistence of a vegetation community other than the climax is the result of interference either by man or by grazing animals. It is perhaps significant that the commonest shrubs which colonize open ground – *Rubus, Crataegus* and *Rosa* – all have armature well adapted to discourage interference by grazing animals or anything else. It may also be significant that the three species described in this book which have straight, or nearly straight prickles, not adapted to assist in climbing, are all erect and self-supporting shrubs characteristic of open habitats.

It can easily be imagined how a grazing animal attempting to tackle *R. pimpinellifolia* would come away with a mouthful of the readily detachable prickles, and anyone who has played golf on a seaside golf course will remember ruefully the occasions when he has sent his ball into the rough consisting of dense impenetrable masses of this rose. However this argument cannot be carried too far, because *R. rubiginosa,* an erect shrub lacking the climbing habit and characteristic of open habitats, has very fiercely hooked prickles.

In south east England and south central England the most extensive open habitat where roses occur in fair abundance is chalk grassland. This habitat was formerly maintained as such by grazing sheep, and more recently after the decline of sheep farming, by rabbits. These animals effectively prevented the development of sapling trees and the more palatable shrubs. Shrubs with protective armature, including the roses, would survive if lucky enough to come through the earlier and more tender stages, giving rise to the areas of open scrub characteristic of this type of grassland. Shortly after the first drastic epidemic of myxomatosis there was a rapid and equally drastic invasion of scrub. At first this considerably added to the interest of these regions for the rhodologist, but if left undisturbed this scrub soon becomes impenetrable as it progresses towards the ultimate climax of woodland, and the roses are eventually shaded out.

In other parts of the country and in Ireland similar areas of grassland maintained by grazing animals are to be found on oolitic, carboniferous or magnesian limestone. Where the landscape is rugged this type of habitat seems likely to be stable; otherwise, as on the chalk, suitable habitats for roses are declining, either for the same reasons as those described for the chalk, or because more and more land is going under the plough. Common land throughout the country was at one time a fruitful hunting ground for roses. In Wolley-Dod's herbarium, collected in the early years of the twentieth century, there are specimens from Ham Common in Surrey and Minchinhampton Common in Gloucestershire of nearly all the southern *Rosa* species. Nowadays urban and other developments and the fact that few people now exercise the right of free grazing on commons has considerably reduced the interest of these areas, especially since many hedgerows have been removed and many of those remaining are mutilated.

The best habitats for roses are now manifestly man-made, such as coal mine and quarry spoil-heaps, worked-out quarries and dismantled railways. Naturally these habitats are by no means permanent and show the same types of succession already described. In the days of steam trains development of scrub on railway verges was effectively controlled by fire, either accidentally caused by sparks from the engines or deliberately started to prevent disastrous accidental ones. Immediately after the Beeching axe the miles of dismantled railway became a paradise for the botanist, exhibiting an enormous variety of herbaceous plants. Some lengths of these were joyfully acquired as nature reserves, but it soon became apparent that most of them were impossible to manage effectively. Inevitably these sites began to be invaded by shrubs, including roses, but the progression towards impenetrable thickets was equally inevitable, along with a subsequent decline in their interest for the rhodologist.

Another frequent habitat for roses is found on pieces of ground where the contours or other factors have prevented ploughing, and which are habitually used for rough grazing. Such places are quite frequent in the Midlands, and the periodic trampling and grazing by cattle or sheep provide ideal conditions for the persistence of the type of open scrub already described. In regions where *R. arvensis* occurs in the nearby hedgerows or woodland this species is found encircling the bases of the larger shrubs such as hawthorn, but rarely on its own in the open. It may be that it prefers the partial shade afforded by the shrubs; the leaf litter below them may provide a closer approximation to woodland soil, or the young rose bushes may be protected from grazing animals.

The distribution maps show the general geographical distribution of the *Rosa* species. However, it should be understood that, for reasons stated elsewhere, these maps are incomplete and uneven in their coverage. There is also the problem of distinguishing between the distribution of a species as defined by climatic or other conditions arising from the geographical location and the distribution of habitats suitable for that particular species. A little thought will show that these are not quite the same thing. *R. pimpinellifolia* illustrates what is meant. It occurs in greatest abundance near the sea coast, its principal habitats there being fixed sand dunes and other suitable areas such as the serpentine heath on the Lizard Peninsula. It also occurs inland with varying frequency in grassland and scrub woodland on chalk and limestone. The map showing the distribution of this species is also showing the distribution of suitable habitats for it. *R. rubiginosa* may be another case in point. There does not appear to be any fundamental geographical limit to the distribution of this species where suitable soils occur, and it may be that the concentration of records in the southern parts of the country merely reflects its preference for the soils on chalk and oolite.

Nevertheless there are species of *Rosa* which do show a definite northerly or southerly distribution. *R. mollis* is a clear-cut northern species, extending southwards down the high ground in the centre of England as far as the

Peak District in Derbyshire, but being almost totally absent except as a rare introduction further south. *R. arvensis* is a southern species, becoming rare in the north of England and extremely rare in Scotland. *R. stylosa* is confined to the south of England, but is quite common in Ireland.

R. canina and the two subspecies of *R. caesia* show distributions which are interesting for another reason. *R. canina* becomes progressively less frequent in the extreme north, and in the north of Scotland is almost completely replaced by *R. caesia*. Similarly *R. caesia* becomes less common, and *R. canina* more common, as one proceed southwards. Again the two subspecies of *R. caesia* show distinct preferences both in habitat and geographical range. Subsp. *glauca* extends to higher altitudes than subsp. *caesia*, both in Scotland and in northern England, and in the extreme north of Scotland and in the Hebrides subsp. *glauca* almost completely replaces subsp. *caesia*. Whilst some authorities would recognize here a single variable species, the altitudinal and geographical preferences of these two very similar taxa provide part of the justification for separating them at subspecific rank.

There is no clear-cut limit to the geographical range of either *R. caesia* or *R. canina*. In the Midlands and somewhat north of them *R. canina* is very common, and, as already stated, *R. caesia* becomes progressively less common southwards. But the hybrids of *R. caesia* and *R. canina* are also frequent in these regions, seemingly out of proportion to the frequency of *R. caesia* itself. One wonders whether in former times *R. caesia* extended in greater frequency further south, and whether climatic changes have favoured *R. canina* at the expense of *R. caesia*. This would account for the present frequency of the hybrids.

Though specimens of *R. caesia* from the more southern regions can definitely be said to belong to this species, they frequently show its distinguishing characters in a somewhat modified or reduced form. Really classical specimens must be looked for in the north. A somewhat similar phenomenon can be observed with *R. sherardii*. This species is recorded from all over the British mainland, but it is commonest in Wales, Scotland and the north of England. When presented with specimens of *R. sherardii* an experienced rhodologist can hazard a very good guess as to where the specimens came from, saying perhaps that one came from Wales, one from Scotland and one from the English Midlands. This is a good example of the local forms which can be found with varying degrees of distinctiveness in every species of the caninoid roses. With the propensity for hybridization in this genus it is inevitable, as explained elsewhere, that allowance must be made for some introgression when deciding what are the permissible limits of variation of a species. Each region has its own particular assemblage of species, and hence the variants of each species are reflections of the local gene pool. Thus in Scotland *R. sherardii* may well be affected by the introgression of *R. mollis,* in Wales of *R. canina*, and in parts of north west England, notably Lancashire, of *R. tomentosa*. This is one of the things which at first tend to puzzle and confuse anyone coming new to the study of *Rosa*. It is essential to become familiar with the roses of one's own region. The experience thus gained will go a long way to dispelling the difficulties first encountered.

The alien species described in this book do not in general have any great effect on the local gene pool described above, probably because they rarely become established in sufficient quantity in any one locality. The commonest of these aliens is undoubtedly *R. rugosa*, which is not only cultivated in gardens, but also used quite widely in landscaping schemes, particularly in coastal regions. From such situations it may escape by bird-sown seed, and once established in suitable habitats may propagate itself vegetatively by suckering. Coastal fixed sand dunes are a frequent habitat for it.

More significant in its effect upon local *Rosa* populations is the presence of garden cultivars of native British species which are not normally found wild in a particular locality. Specimens found as escapes in the wild can often be recognized by their characters as cultivars. Their presence in gardens and as occasional escapes does, however, provide a source of pollen which can introduce the genes of these species into the local pool. The principal species responsible for this are *R. pimpinellifolia* and *R. rubiginosa.* That the presence of *R. rubiginosa* in parts of Wales appears to be derived from garden stock is testified by personal communications from A.O. Chater and Miss A.P. Conolly who note it in the Teifi valley of Cardiganshire and in the Lleyn Peninsula respectively.

Just how effective animals, including migrating birds, can be as a means of dispersing native species into regions beyond their normal range is a matter for debate. Some seeds, such as those of hawthorn, bramble, strawberry and gooseberry are eaten by the animal along with the succulent parts of the fruit and pass unharmed through the alimentary canal. In this way they can be carried considerable distances before they are finally deposited in a place where they can germinate. However there is evidence that birds and probably other animals eat the succulent tissues of the rose hip and reject the achenes. In many species the achenes are covered with distasteful and irritant hairs – schoolboys in rural districts are often aware that when collected and dried these hairs make an excellent itching powder. It is doubtful whether birds or other animals would carry whole hips any great distance before eating the hip tissue and rejecting the achenes. A much more likely reason for the presence of species outside their normal range is the recent practice adopted by local authorities during landscaping of planting 'wild' roses obtained from nurserymen. These are often an extraordinary and incongruous assemblage of species, many of them garden cultivars or species quite alien to the soils and localities in which they are planted. Also *R. rugosa, R. canina* and perhaps other wild species are used by nurserymen as stocks on to which various cultivars are grafted. These, especially if they have a suckering habit, as does *R. rugosa,* may grow up round the base of the bush, and sometimes may oust the scion which was grafted on to them. These rogue bushes, together with the supplies of suitable bushes kept by nurserymen for grafting stocks, may result in the escape of such taxa into the wild, or the introduction of their genes into the local pool.

6. On collecting and pressing roses

We can perhaps best begin by quoting some words of warning given by two British rhodologists over eighty years ago (Ley & Wolley-Dod 1909) "The determination of specimens is often much hampered by their inadequacy. They are, as a general rule, collected too young. The size and colour of the flowers are all that is lost by a late collection and although these have their importance, they are far less indispensable than fully-developed fruit, without which it is often not easy to determine even the group to which the sample should be referred. Specimens are also very commonly too small. The end of a flowering shoot rarely shows the characteristic prickles, which are to be found on the barren shoots; or on the old stems from which the flowering-shoots grow. Either of these is admissible, but the very strong shoots of the year, arising from the rootstock, should be avoided, because their strength gives deceptive characters to the prickles and leaves which they bear."

If one had to deal only with good species many of the problems, dealt with above, might not arise. However as rhodologists now recognise a multiplicity of hybrids, back crosses and introgressions – some of which are far from rare – as well as the dozen or so native species, the collecting of specimens is extremely important.

Collecting may be for one or more of the following purposes:

1. Collecting from a single bush in order to send to a referee and thus obtain a name.
2. Collecting from a wide area during a survey in order to have the more difficult specimens named, the collector already having some familiarity with the common species of the area.
3. Collecting with the intention of getting to know the taxa in one's own area.
4. Collecting representative specimens for a herbarium.

The procedure to be followed differs slightly in each case.

(1.) is to be discouraged unless the record is needed for an ecological survey and the collector has no time in which to pursue his investigations further. In this case a sufficiency of material should be collected and sent, whilst fresh, to a referee.

(2. & 3.) Collecting should be done intelligently after perusing the following notes.

(4.) The amount of material will of necessity be limited by the size of the herbarium sheet. A selection should therefore only be made after studying the following notes and possibly also the morphology chapter.

Minimum requirements for a collection.

In general a couple of fruiting sprays, collected when the hips are well-developed, though not necessarily fully ripe, together with a piece of second-year stem to which one or two representative leafy sprays are attached, form the minimum requirements for a diagnosis. Further evidence should be collected in accordance with the following notes, particularly if the plant is thought to be a hybrid. When bushes grow intermingled, it is important to ensure that the material collected is not mixed.

A few notes on certain aspects of the plant, which cannot be represented in a specimen or on a herbarium sheet, may also prove to be invaluable or even indispensable.

General notes on collecting material.

Habit, colour etc.

In *Rosa* much can be deduced from the habit and general appearance of a bush. Height, strength, presence of trailing branches and length of internodes may be impossible to deduce from a specimen and even the contrasting colours of stem and leaves may not be fully displayed in the material selected.

Prickles.

Their number, size, shape and range of variation are important and every attempt should be made to display the range of variation. Mixed armature is especially important and it may be necessary, after a thorough examination of a bush, to include more than one portion of stem. The stems can be split longitudinally for mounting purposes or, alternatively, slivers of stem with prickles attached may be substituted on a herbarium sheet. In the *Rubiginosae* the presence or absence of fine acicles is important and, as their distribution is not uniform, it is important to include them in any specimen or to note their entire absence.

Leaflets.

Their number, clothing and dentition are usually adequately represented in a specimen, but as leaflets may differ in shape and size on any bush care should be taken to obtain a representative sample. In hybrids the influence of the male parent may be more pronounced towards the base or apex of a shoot. Such variation, if present, should be noted or sampled.

Hips.

Hip shape is often liable to considerable variation, even on the same bush, the central, or primordial hip of a cluster being often more elongate than the rest. As with other organs a representative sample should be selected. An exception to this rule may be made when shrivelled or aborted hips are present. As this phenomenon may be indicative of hybridity it should be noted and if possible sampled.

Pedicels.

The presence or absence of glandular setae or acicles on the pedicels is important in the separation of rose taxa. Their presence in small quantity on a specimen with normally eglandular pedicels will indicate introgression or hybridity. Examples should be included in the material collected.

Sepals.

The degree of pinnation and clothing of sepals is important. If not attached to the hips, or if disarticulating when collected, a few sepals should be put in a packet for reference. The way the sepals are held on the ripening fruit is of extreme importance as well as their time of disarticulation. This is a dynamic process which can be arrested and prolonged in certain species by collecting hips at the appropriate time. However in pressed and dried specimens these characters are often altered or obscured. Notes should therefore be made of the dynamics of the sepals as the fruit matures. This information can often be obtained in one visit as hips in various stages of development are usually to be found on the one bush.

Notes to be appended to specimen or herbarium sheet.

As indicated above, any features which cannot be portrayed in a specimen should be carefully noted and appended to the specimen or to the herbarium sheet. These should include:

1. The form and aspect of the bush, the direction and strength of its internodes, whether suckering or not and if possible its height.
2. Any colouring of stem or leaves.
3. Position of sepals on maturing hips.
4. Colour and size of flowers if noted on a previous or later visit. Such notes are useful, if not for a first determination, at least for the record and for further study.
5. Location (including grid reference), date, collector.

Conclusion.

It will be seen that informed and selective sampling will not only help in making a correct diagnosis of a plant but will also help in reducing the material collected to a minimum.

Beginners should however collect too much rather than too little material. Pruning of rose bushes in the course of collection is unlikely to do harm, just as horticultural pruning does no harm to the cultivars.

Pressing roses.

The main difficulty in pressing roses lies in treating extremely thick structures such as stems and hips, which do not or should not yield to moderate pressure, along with leaves, sepals, etc., which shrivel when not enough pressure is applied.

Sufficient packing should be applied to the thin structures to prevent their shrivelling in the drying process. This is best carried out over a low heat. When no heat is applied drying papers should be changed several times.

One or two loose hips and a few sepals may be included in a packet appended to the appropriate herbarium sheet. By this means stigmas, styles and stylar aperture can be examined in the future without breaking hips from the main specimen.

"Most roses are in the best condition for determination [and pressing] from the last week in July till the end of September, and often later so long as the leaves have not fallen." (Wolley-Dod 1924). This depends, of course, on locality and latitude !

7. Classification

The family Rosaceae contains many genera which are used for ornament, for their attractive scent and for food. Hutchinson (1964) lists 124 genera and gives the species complement as 3375.

The genera, and *Rosa* in particular, are mostly found in North-temperate and sub-tropical zones. According to Krüssmann (1981) there are no indigenous roses in the Southern hemisphere. We are concerned with a single genus and the species thereof that occur in the wild and are native or naturalized in Britain and Ireland.

Our garden roses are hybrids of complex origin, derived from a number of species, and have been developed over a very long period of time. Several foreign species are also fairly often grown. Those which escape into the wild are often cultivars, differing considerably from the species or hybrids from which they derive. Little is yet known about which taxa are naturalized in Britain and Ireland, and even less about their distribution here. The keys, descriptions and distributions in this Handbook should for the most part be considered as provisional.

Rosa L. (after Clapham, Tutin & Moore 1987)

Shrubs, sometimes trailing or scrambling, usually deciduous. Leaves pinnate (very rarely simple); stipules (usually) adnate to the petiole. Stems usually prickly. Flowers terminal, solitary or in corymbs, hermaphrodite, (4-)5-merous. Stamens numerous. Styles protruding through the orifice of a disc. Ovule 1, pendulous. Flowers homogamous, mostly without nectar, visited by various insects for pollen, self-pollination possible if insect visits fail. Fruit of numerous achenes enclosed in the coloured fleshy receptacle ('hip').

8. Synopsis of classification

There is no completely satisfactory classification as few botanists possess the wide knowledge necessary for the task. Déséglise (1864-65) is often followed. He considers the numerous classifications which have been adopted between those of Linnaeus in 1764 and of Reuter in 1861 and concludes his article with a 'Synopsis Specierum', but notes that "The classification which I propose, is not, I know, more than any other, beyond the reach of criticism; and I must ask my readers.....to consider it only as a fresh attempt to facilitate our knowledge of our French species."

As we are dealing with only a small fraction of the European roses we shall not attempt to place the few British species within Déséglise's classification but follow the order of that of the most recent British Flora (Stace 1991).

Alien species are marked with an asterisk (*). Chromosome counts of all the other species have been made on British material.

Section SYNSTYLAE DC.
   1. *R. multiflora Thunb. ex Murr.   Diploid $2n = 14$
   2. *R. setigera Michaux   Diploid $2n = 14$
   3. *R. luciae Franchet & Rochebr.   Diploid $2n = 14$
   4. R. arvensis Hudson   Diploid $2n = 14$

Section PIMPINELLIFOLIAE DC.
   5. R. pimpinellifolia L.   Tetraploid $2n = 28$

Section CINNAMOMEAE Ser. (Subgen. *Cassiorhodon* Dumort.)
   6. *R. rugosa Thunb. ex Murr.   Diploid $2n = 14$
   7. *R. 'Hollandica'
   8. *R. glauca Pourret   Tetraploid $2n = 28$

Section CAROLINAE Crépin
   9. *R. virginiana Herrm.   Tetraploid $2n = 28$

Section ROSA (Section *Gallicanae* DC.)
   10. *R. gallica L.   Tetraploid $2n = 28$

Section CANINAE DC.

  Subsection STYLOSAE Crépin
   11. R. stylosa Desv.   Unbalanced $2n = 35$ or 42

  Subsection CANINAE Crépin
   12. R. canina L.   Unbalanced $2n = 35$
   13. R. caesia Sm.   Unbalanced $2n = 35$
     13a. Subsp. caesia   Unbalanced $2n = 35$
     13b. Subsp. glauca (Nyman)
         G.G. Graham & Primavesi   Unbalanced $2n = 35$
   14. R. obtusifolia Desv.   Unbalanced $2n = 35$

Subsection VILLOSAE (DC.) Crépin
    15. *R. tomentosa* Sm.   Unbalanced $2n = 35$
    16. *R. sherardii* Davies   Unbalanced $2n = 28, 35$ or 42
    17. *R. mollis* Sm.   Unbalanced $2n = 28, 56$

Subsection RUBIGINOSAE Crépin
    18. *R. rubiginosa* L.   Unbalanced $2n = 35$
    19. *R. micrantha* Borrer ex Sm.   Unbalanced $2n = 35$ or 42
    20. *R. agrestis* Savi   Unbalanced $2n = 35$ or 42

9. List of species and hybrids noted in the text

In the list, reciprocal hybrids are recognized, the female parent being listed first in each case. Hybrids enclosed in [ ] are doubtfully recorded. Taxa marked with an asterisk are not native in Britain and Ireland.

1	*R. multiflora* Thunb. ex Murray	
2	*R. setigera* Michaux	
3	*R. luciae* Franchet & Rochebr.	
4	R. arvensis Huds.	
4 × 10	*R. arvensis × gallica	R. × alba L. (R. × collina Jacq.)
4 × 11	R. arvensis × stylosa	R. × pseudorusticana Crépin ex Rogers
4 × 12	R. arvensis × canina	R. × verticillacantha Mérat
4 × 15	R. arvensis × tomentosa	
4 × 16	R. arvensis × sherardii	
4 × 18	R. arvensis × rubiginosa	R. × consanguinea Gren.
[4 × 19	R. arvensis × micrantha]	[R. × inelegans W.-Dod]
5	R. pimpinellifolia L.	
5 × 12	R. pimpinellifolia × canina	R. × hibernica Templeton
5 × 13	R. pimpinellifolia × caesia	R. × margerisonii (W.-Dod) W.-Dod
5 × 15	R. pimpinellifolia × tomentosa	R. × coronata Crépin ex Reuter
5 × 16	R. pimpinellifolia × sherardii	R. × involuta Sm.
5 × 17	R. pimpinellifolia × mollis	R. × sabinii Woods
5 × 18	R. pimpinellifolia × rubiginosa	R. × cantiana (W.-Dod) W.-Dod
6	*R. rugosa Thunb. ex Murray	
6 × 12	*R. rugosa × canina	R. × praegeri W.-Dod
7	*R. 'Hollandica'	
8	*R. glauca Pourret (R. rubrifolia Villars, nom. illegit.)	
9	*R. virginiana Herrm.	
10	*R. gallica L.	
11	R. stylosa Desv.	
11 × 4	R. stylosa × arvensis	R. × pseudorusticana Crépin ex Rogers
11 × 12	R. stylosa × canina	R. × andegavensis Bast.
11 × 13	R. stylosa × caesia	
11 × 20	R. stylosa × agrestis	
12	R. canina L.	
12 × 4	R. canina × arvensis	R. × verticillacantha Mérat
12 × 5	R. canina × pimpinellifolia	R. × hibernica Templeton
12 × 11	R. canina × stylosa	R. × andegavensis Bast.
12 × 13	R. canina × caesia	R. × dumalis Bechst.
12 × 14	R. canina × obtusifolia	R. × dumetorum Thuill.

47

12 × 15	R. canina × tomentosa	R. × scabriuscula Sm.
12 × 16	R. canina × sherardii	
12 × 17	R. canina × mollis	R. × molletorum H.-Harr.
12 × 18	R. canina × rubiginosa	R. × nitidula Besser
13	R. caesia Sm.	
13a	Subsp. caesia	
13b	Subsp. glauca (Nyman) G.G. Graham & Primavesi	
13 × 4	R. caesia × arvensis	
13 × 5	R. caesia × pimpinellifolia	R. × margerisonii (W.-Dod) W.-Dod
13 × 12	R. caesia × canina	R. × dumalis Bechst.
13 × 15	R. caesia × tomentosa	R. × rogersii W.-Dod
13 × 16	R. caesia × sherardii	
13 × 17	R. caesia × mollis	R. × glaucoides W.-Dod
13 × 18	R. caesia × rubiginosa	
13 × 19	R. caesia × micrantha	R. × longicolla Ravaud ex Rouy
14	R. obtusifolia Desv.	
14 × 4	R. obtusifolia × arvensis	R. × rouyana Duffort ex Rouy
14 × 11	R. obtusifolia × stylosa	
14 × 12	R. obtusifolia × canina	R. × dumetorum Thuill.
14 × 13	R. obtusifolia × caesia	
14 × 15	R. obtusifolia × tomentosa	
14 × 18	R. obtusifolia × rubiginosa	R. × tomentelliformis W.-Dod
15	R. tomentosa Sm.	
15 × 5	R. tomentosa × pimpinellifolia	R. × coronata Crépin ex Reuter
15 × 12	R. tomentosa × canina	R. × scabriuscula Sm.
15 × 13	R. tomentosa × caesia	R. × rogersii W.-Dod
15 × 14	R. tomentosa × obtusifolia	
15 × 16	R. tomentosa × sherardii	R. × suberectiformis W.-Dod
15 × 17	R. tomentosa × mollis	
15 × 18	R. tomentosa × rubiginosa	R. × avrayensis Rouy
15 × 19	R. tomentosa × micrantha	
15 × 20	R. tomentosa × agrestis	
16	R. sherardii Davies	
16 × 4	R. sherardii × arvensis	
16 × 5	R. sherardii × pimpinellifolia	R. × involuta Sm.
16 × 12	R. sherardii × canina	
16 × 13	R. sherardii × caesia	
16 × 15	R. sherardii × tomentosa	R. × suberectiformis W.-Dod
16 × 17	R. sherardii × mollis	R. × shoolbredii W.-Dod
16 × 18	R. sherardii × rubiginosa	R. × suberecta (Woods) Ley
16 × 19	R. sherardii × micrantha	
16 × 20	R. sherardii × agrestis	
17	R. mollis Sm.	

17 × 5	*R. mollis* × *pimpinellifolia*	*R.* × *sabinii* Woods
17 × 12	*R. mollis* × *canina*	*R.* × *molletorum* H.-Harr.
17 × 13	*R. mollis* × *caesia*	*R.* × *glaucoides* W.-Dod
17 × 16	*R. mollis* × *sherardii*	*R.* × *shoolbredii* W.-Dod
17 × 18	*R. mollis* × *rubiginosa*	*R.* × *molliformis* W.-Dod
18	*R. rubiginosa* L.	
18 × 4	*R. rubiginosa* × *arvensis*	*R.* × *consanguinea* Gren.
18 × 5	*R. rubiginosa* × *pimpinellifolia*	*R.* × *cantiana* (W.-Dod) W.-Dod
18 × 11	*R. rubiginosa* × *stylosa*	
18 × 12	*R. rubiginosa* × *canina*	*R.* × *nitidula* Besser
18 × 15	*R. rubiginosa* × *tomentosa*	*R.* × *avrayensis* Rouy
18 × 16	*R. rubiginosa* × *sherardii*	*R.* × *suberecta* (Woods) Ley
18 × 17	*R. rubiginosa* × *mollis*	*R.* × *molliformis* W.-Dod
19	*R. micrantha* Borrer ex Sm.	
19 × 4	*R. micrantha* × *arvensis*	*R.* × *inelegans* W.-Dod
[19 × 11	*R. micrantha* × *stylosa*]	
19 × 12	*R. micrantha* × *canina*	*R.* × *toddiae* W.-Dod
19 × 14	*R. micrantha* × *obtusifolia*	
19 × 18	*R. micrantha* × *rubiginosa*	*R.* × *bigeneris* Duffort ex Rouy
19 × 20	*R. micrantha* × *agrestis*	*R.* × *bishopii* W.-Dod
20	*R. agrestis* Savi	
20 × 11	*R. agrestis* × *stylosa*	
20 × 12	*R. agrestis* × *canina*	*R.* × *belnensis* Ozan
[20 × 16	*R. agrestis* × *sherardii*]	
[20 × 19	*R. agrestis* × *micrantha*]	*R.* × *bishopii* W.-Dod

R. elliptica Tausch has been erroneously recorded.
The name *R. villosa* L. has been misapplied to British plants.
R. caesia × *rugosa* and *R. mollis* × *rugosa* need careful redetermination.

10. Keys

The determination derived from any key should always be checked by careful comparison of the specimen with the account of the species in the systematic section. There are occasional anomalies which would render these or any other keys misleading. The following are the most likely sources of confusion:

1. *R. stylosa* occasionally has glabrous leaves. Also the pedicels are occasionally smooth, or the glands tend to fall off.
2. There is a variety of *R. canina*, common in the Midlands, with a conical disc.
3. If no satisfactory conclusion can be reached by using the keys, or if the specimen does not coincide with any of the species accounts, then it is almost certainly a hybrid (or just possibly a garden escape not included in this Handbook).

Dichotomous key to native and alien species

Note: Two hybrids are included in this key, because (a) they occur frequently in the absence of at least one parent; (b) they are easily recognized; and (c) omission of them from the key might cause confusion in the field.

1. Leaves glabrous (occasionally with a few sparse hairs on midrib beneath) 2
1. Leaves conspicuously hairy, at least on the veins beneath 10

2. Styles fused together in a long-exserted column 3
2. Styles free and not long-exserted 5

3. Climbing plant; leaflets very large, usually 3 (rare alien) **2. R. setigera**
3. Weakly trailing or procumbent; leaflets smaller, 5 or more 4

4. Styles pubescent (rare alien or escape) **3. R. luciae**
4. Styles glabrous (common except in the north) **4. R. arvensis**

5. Sepals entire; erect free-standing shrub 6
5. Sepals lobed; climbing plant with arching stems 8

6. Low erect shrub c. 50cm tall, suckering, with numerous, slender prickles and acicles; leaflets (7-)9-11, c.1cm, small; hips purplish-black when ripe **5. R. pimpinellifolia**
6. Taller; prickles more robust; leaflets fewer than 9; hips red when ripe (rare alien or escape) 7

7. Leaflets green **9. R. virginiana**
7. Leaflets with strong reddish tinge **8. R. glauca**

8. Pedicels variable in length but not hidden by bracts; stigmas in a small globose head, not covering the disc **12. R. canina**
8. Pedicels short, partly hidden by large bracts; stigmas in a domed villous head, partly or completely covering the disc   9

9. Sepals erect or spreading-erect, persistent until the hips are ripe
 13b. R. caesia subsp. **glauca**
9. Sepals reflexed, falling before the hips are ripe
 13b × 12. R. caesia subsp. **glauca × R. canina**

10. Disc strongly conical; styles connivent at first in a short exserted column, later becoming separated **11. R. stylosa**
10. Disc not conical, styles not exserted, or exserted and permanently fused into a column   11

11. Subfoliar glands numerous, conspicuous, viscid, with fruity odour   12
11. Subfoliar glands absent, or with resinous odour or odourless   14

12. Styles hispid; sepals spreading-erect, persistent until the hips are ripe; pedicels c. 1cm, short, glandular-hispid; erect shrub with straight stems **18. R. rubiginosa**
12. Styles glabrous; sepals reflexed, falling before the hips are ripe; pedicels usually more than 1 cm, glandular-hispid or not; young stems arching or flexuous   13

13. Climbing plant with arching stems; pedicels glandular-hispid; leaflets rounded at base **19. R. micrantha**
13. Free-standing shrub with somewhat flexuous young stems; pedicels smooth; leaflets cuneate at base **20. R. agrestis**

14. Styles united in a long-exserted column; flowers and hips numerous, in clusters of 10 or more (rare alien or escape)
 1. R. multiflora
14. Styles not exserted; flowers and hips less numerous   15

15. Stems tomentose, with numerous straight prickles and acicles; hips 2cm or more, large, globose or depressed-globose; pedicels curved (alien often planted) **6. R. rugosa**
15. Acicles absent; hips usually less than 2cm; pedicels straight   16

16. Leaflets uniserrate, with subfoliar glands; sepals c.10 × 5mm,
    short and broad (rare alien or escape) **10. R. gallica**
16. Leaflets multiserrate, biserrate or uniserrate but, if uniserrate,
    without subfoliar glands; sepals more than 10mm 17

17. Pedicels smooth; leaflets pubescent but not tomentose 18
17. Pedicels glandular-hispid; leaflets tomentose 21

18. Pedicels short, partly concealed by large bracts; stigmas in a
    domed villous head, partly or completely concealing the disc 19
18. Pedicels long or short, not concealed by bracts; stigmas not
    covering the disc 20

19. Sepals erect or spreading-erect, persistent until the hips are ripe
    13a. R. caesia subsp. **caesia**
19. Sepals reflexed, falling before the hips are ripe
    13a × 12. R. caesia subsp. **caesia × R. canina**

20. Leaflets biserrate with dark reddish-brown glands on teeth;
    sepals bipinnate, strongly reflexed and pressed to the hip;
    pedicels 0.5-1.5cm **14. R. obtusifolia**
20. Leaflets usually uniserrate and eglandular; sepals loosely reflexed,
    pinnate; pedicels usually 1.5-2.5cm **12. R. canina** Group **Pubescentes**

21. Climbing plant with arching stems; sepals spreading or spreading-erect
    but falling before the hips are ripe; pedicels 2-3.5cm;
    stylar orifice 1/5 diameter of the disc or less **15. R. tomentosa**
21. Erect shrub; sepals erect, persisting at least until the hips are ripe;
    pedicels 0.5-1.5cm; stylar orifice at least 1/3 diameter of the disc 22

22. Prickles curved; stems slightly zig-zag, flexuous towards apex;
    leaflets ashy-grey or bluish-green (sometimes reddish tinged);
    sepals lobed, long-persistent but eventually falling, thin and
    papery at base **16. R. sherardii**
22. Prickles straight; stems straight; leaflets dark-green;
    sepals more or less simple, often red and fleshy at the base,
    persistent until the hips rot **17. R. mollis**

Alternative dichotomous key to native species

Note: In areas where *R. canina* and *R. caesia* overlap the hybrid *R.* × *dumalis* is frequent; it is not included in the following key but should be kept in mind if a specimen appears to match fully neither *R. canina* nor the two subspecies of *R. caesia*

1. Styles more or less fused into a slender column and long-exserted from a small flattish disc; weakly scrambling or trailing shrub   **4. R. arvensis**
1. Styles free, scarcely exserted or, if agglutinated and shortly exserted, the disc strongly conical; climbing or erect shrub   2

2. Styles at first agglutinated into a plump column, protruding from the disc, later separating into individual strands; stigmas in a tiered oval head; disc strongly conical, thick in section   **11. R. stylosa**
2. Styles free, included (may be slightly protruding in dried material), never falling apart but remaining in a closely packed bunch; stigmas in a small globose head or in a hemispherical mass partially or completely covering the disc; disc flat, concave, or, if slightly conical, thin in section   3

3. Stems with dense prickles and acicles; leaflets (7-)9-11, 0.5-1.5(-2)cm, small; flowers solitary, without bracts; hips purplish-black on ripening   **5. R. pimpinellifolia**
3. Prickles more widely-spaced, the acicles (if present) in discrete areas; leaflets 5-7(-9), mostly more than 1cm; flowers several; bracts always present; hips red on ripening   4

4. Leaflets glabrous, sometimes with a few sparse hairs on the midrib beneath   5
4. Leaflets tomentose or hairy beneath, at least on the midrib   6

5. Styles glabrous or hispid, with a small head of stigmas not much bigger than the orifice; stylar orifice c.1/6-1/5 diameter of the disc; pedicels normally more than 1.5 cm; sepals eventually reflexed and falling before the hips ripen; anthocyanin pigmentation weak   **12. R. canina**
5. Stigmas forming a dense, villous, hemispherical mass almost covering the disc; stylar orifice c.1/3 diameter of the disc; pedicels normally less than 1.5cm; sepals spreading-erect to ascending-erect after flowering and persisting until the hips ripen; anthocyanin pigmentation strong except when growing in shade   **13b. R. caesia** subsp. **glauca**

6. Plants with numerous glands; glands viscid, brownish or translucent, evidently stalked, some c.90-100μm in diameter, usually with a strong fruity odour when rubbed 7
6. Plants eglandular, or with few or numerous glands; glands, if present, 50μm or less in diameter, on very short stalks, without a fruity odour when rubbed 9

7. Most leaflets cuneate at base; glands with weak scent; pedicels glabrous, without glands (very rare except in S. Ireland) **20. R. agrestis**
7. Leaflets more or less rounded at base; glands usually with a fruity odour; pedicels glandular-hispid 8

8. Stems erect, strong; prickles unequal; acicles sometimes present on the pedicels or flowering branches; sepals ascending or spreading-erect, persistent until the hips ripen; styles and stigmas hispid **18. R. rubiginosa**
8. Stems weak, climbing and arching; prickles equal; acicles absent; sepals reflexed, falling before the hips ripen; styles and stigmas glabrous **19. R. micrantha**

9. Pedicels glandular-hispid; glands on leaflets numerous, often aromatic, whitish or a glossy red, c.50μm in diameter 10
9. Pedicels smooth; glands absent from the leaflets or, if present, few and restricted to the veins beneath, always odourless, dull red or brownish, 30μm or less in diameter 12

10. Strongly climbing, often to well over 2m; prickles slightly curved but strong; glands odourless; pedicels (2-)2.5-3.5cm, long; stylar orifice c.1/5 diameter of the disc; stigmas glabrous to villous, not covering the disc; sepals spreading to spreading-erect, falling well before the hips ripen **15. R. tomentosa**
10. Stems stiffly or weakly erect, free-standing and not climbing; prickles straight, curved, or slightly curved, weak; glands strongly aromatic when crushed: pedicels mostly less than 2cm; stylar orifice 1/3 diameter of the disc or more; stigmas villous, in a hemispherical head covering the disc; sepals spreading-erect to erect, persisting at least until the hips begin to change colour 11

11. Stems straight, stiffly erect, suckering; prickles straight; sepals erect, more or less simple; pedicels 0.5-1.5cm; stylar orifice 1/2 diameter of the disc **17. R. mollis**

11. Stems erect but weaker and somewhat flexuous; prickles slightly curved, weak (more strongly curved in some areas); sepals spreading-erect, pinnate; pedicels 1-1.5cm; stylar orifice c.1/3 diameter of the disc **16. R. sherardii**

12. Sepals bipinnate with large lobes, strongly reflexed after flowering, falling before the hips ripen; stigmas in a small, globose, hispid head; stylar orifice 1/5 diameter of the disc or less **14. R. obtusifolia**
12. Sepals pinnate, spreading-erect to ascending-erect after flowering, persisting until the hips ripen; stigmas in a large, villous, hemispherical head concealing the disc; stylar orifice 1/3 diameter of the disc **13a. R. caesia** subsp. **caesia**

11. Description and figures

Notes on descriptions

The range of variation in measurements and numbers is of necessity approximate. Because of environmental and other factors anomalies may be found. Measurements are length, or length × width unless otherwise stated.

The description of prickles refers to those on the more mature, but not oldest, parts of the stem. Prickles on the oldest wood and on young leading shoots are often not typical.

Leaflets described as glabrous may sometimes have a very few sparse hairs on the lower surfaces but never regularly on veins or midrib.

Numbers of flowers refer to the number in each inflorescence.

Descriptions and measurements of hips are of those which have reached their ultimate mature size. The hips need not necessariy be ripe, as maturity of size is often reached whilst the hips are green.

The stylar orifice of most species is not a perfect cylinder but expands somewhat as the disc is reached. For a comparison with the diameter of the disc the orifice must be measured or estimated at its narrowest bore. Styles should be pulled out first and the hip viewed from above. Sections can easily be made on fresh hips.

The descriptions of hybrids should be taken as a guide only. The ability to recognize hybrids is only acquired after becoming familiar with the characters of the species. It should also be noted that some of the rarer hybrids are described from single specimens or, occasionally, from inadequate herbarium material.

After each hybrid description a list of Watsonian Vice-Counties in which that particular hybrid has been found is appended. It should be understood that many hybrids are under-recorded and that this list represents only the present state of our knowledge.

Note on Figures

The species figures each comprise a portion of fruiting stem with typical leaves, a piece of mature stem with prickles, details of lower surface of leaflet showing serration, pubescence and glands, hips showing details of disc, orifice and styles, and a median vertical section of a hip. Scale bars are appended to each item, and in each case represent 1cm.

1. Rosa multiflora Thunb. ex Murray

Many-flowered Rose

Climbing shrub 3-5m, very strongly growing, densely branched; stems moderately prickly, sometimes nearly unarmed; prickles curved and small. Leaflets (5-)7, 2-3 × 1.2-1.8cm, ovate or narrowly so, shallowly crenate-serrate, dull matt green above, lighter green and sometimes pubescent beneath; petiole and rachis rough with short glands and pricklets; stipules 15 × 1mm, very narrow, glandular, minutely pinnate. Flowers more than 10, 2-3cm in diameter, in conical corymbs, white; petals narrow and not much overlapping. Hips 7mm, small, ovoid to globose, red when ripe; pedicels 1-1.5cm, glandular-hispid, glands very short; sepals 5 × 4mm, triangular, shorter than the petals even in bud, falling early; styles united in a glabrous column and well-exserted; disc 2-3mm; orifice minute. $2n = 14$. Flowering June – August.

Useful in dense hedges and as an understock for budding. Naturalized in hedges and woods in scattered parts of England and south-west Scotland. Native of East Asia.

V.-c. 1,2,16,17,28,29,64,73.

R. multiflora 1

2. Rosa setigera Michaux

Prairie Rose

Climbing or trailing shrub 1-2(-5)m, with several, slender stems reaching 3-4m in length in a season; stems nearly unarmed, occasionally with a few small, curved prickles, young branches unarmed. Leaflets 3(-5), 3-9 × 4-5cm, very large, ovate or oblong, crenate-serrate, somewhat rugose, dark green above, greyish-green and glabrous or sparsely hairy on the veins beneath; petiole and rachis rough with short stipitate glands and pricklets; stipules 25 × 1.5mm, very narrow, minutely gland-fringed, and with slender spreading auricles. Flowers 4-10, 4-6cm in diameter, white to dark pink, in lax corymbs. Hips 0.5-0.8cm, small, ovoid, minutely glandular-hispid, brownish green; pedicels 1.5-2.5cm, glandular hispid, the glands very short; sepals 15 × 5mm, entire or with a few linear appendages, minutely glandular, reflexed and deciduous; styles in a long agglutinated, twisted column, hispid, terminating in a loose cylindrical, hispid group of stigmas; disc 3.5mm, flat; orifice very small. $2n = 14$. Flowering June – August.

Used for breeding hardy climbing roses. Well naturalized in scrub in Jersey and Guernsey (Stace, 1991). Native of North America.

V.-c. CI.

R. setigera 2

61

3. Rosa luciae Franchet & Rochebr.

Memorial Rose

Prostrate, with long, trailing branches or climbing to 3m, in which case the ends of the stems hang vertically downwards; prickles small, slender, curved. Leaflets 7-9, 1.8-2.5 × 1-1.5cm, rather thin, widely obovate or rounded acuminate, simply crenate-serrate, the teeth tipped with rudimentary glands, glabrous, glossy, dark-green above, lighter green, with a few sparse hairs on the midribs beneath; petiole and rachis pubescent, with pricklets, sparsely glandular; stipules 10-15 × 1.5mm, very narrow, glandular-laciniate. Flowers 5-8, 3-5cm in diameter, white, scented, in very irregular corymbs. Hips 6-10 × 4-6mm, small, ovoid or globose, somewhat urceolate, red when ripe; pedicels 1.5-2.5cm, densely glandular-hispid; sepals 6 × 3mm, entire, or with a few appendages, glandular and hispid, reflexed and deciduous before the fruit ripens; styles pubescent in a united, exserted column; disc 3-4mm, flat; orifice very small. $2n = 14$. Flowering May – June.

Because of its prostrate habit it is sometimes used for ground cover. Naturalized in open ground and in scrub, usually near the sea, in Jersey, Guernsey, E. Kent and Dunbarton (Stace 1991). Native of East Asia.

V.-c. 15,99,CI.

R. luciae 3

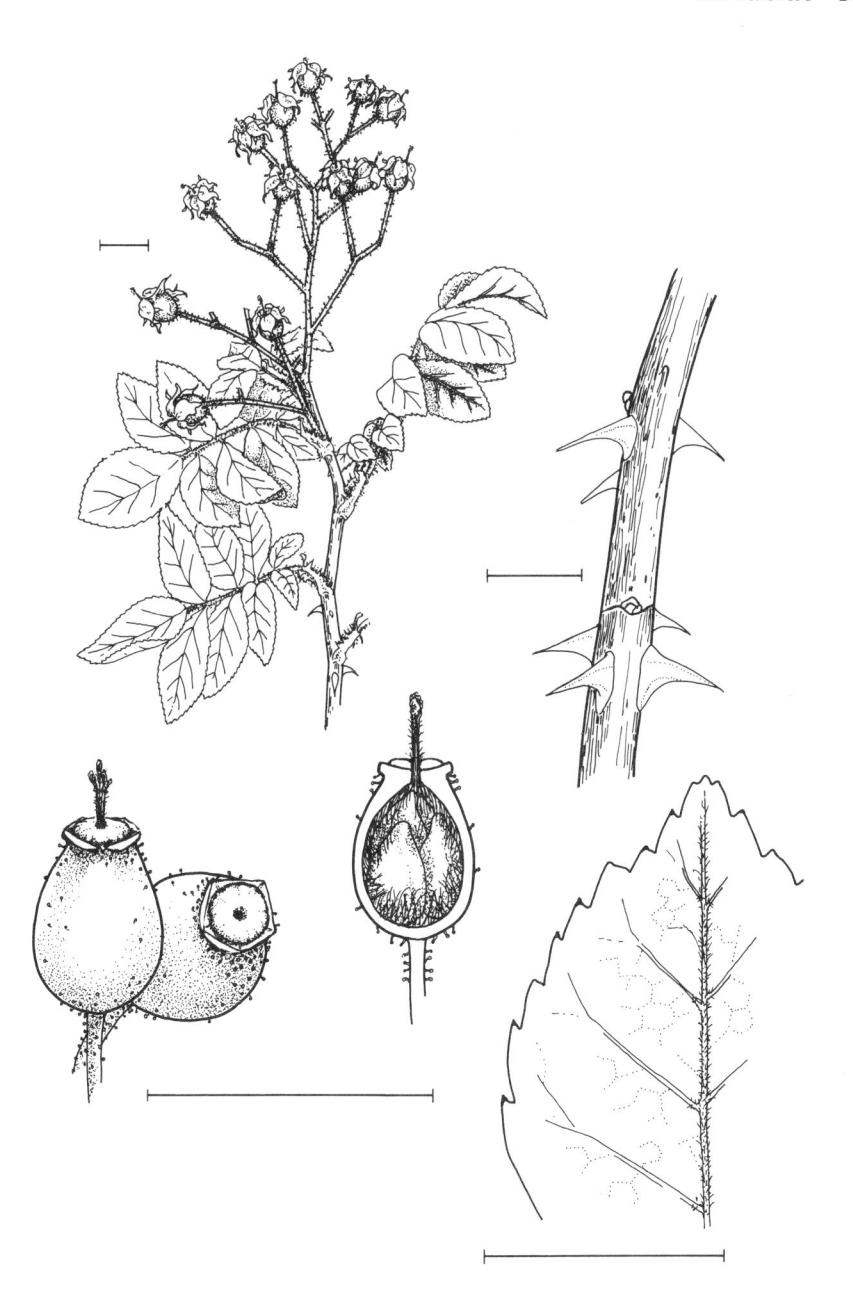

4. Rosa arvensis Huds.   Map 2

Field-rose

Trailing, and forming low, sprawling masses 50-100cm high or climbing over other shrubs; stems weak, green-glaucous with a strongly contrasted wine-red, sunlit side; prickles curved, slender but strong or stout on older stems. Leaflets 5-7, 1-3 × 0.6-1.3cm, ovate-elliptical, crenate-serrate, rather thin, green, paler and glabrous or pubescent on the veins beneath, eglandular; petiole and rachis usually with some stalked glands and pricklets; stipules 12 × 4mm. Flowers 1-6, 3-5cm in diameter, white with cream centre, sharply contrasted with the bright yellow anthers. Hips 0.8-1.5 × 0.5-1cm, narrowly ovoid to globose, red when ripe, smooth; pedicels 2-5cm, long, with short-stalked glands, rarely smooth; sepals less than 1cm, ovate-acuminate, often purplish, simple or with very small lobes, eglandular, falling early; styles glabrous, fused into a slender, exserted column equalling the stamens; stigmas in a small globose head; disc almost flat, thick; orifice 1/6 diameter of disc or less. $2n = 14$. Flowering June – July.

A typical woodland species, tolerant of some shade. It also occurs in hedgerows, especially near woodland and in open scrub where it may form large sprawling masses, often surrounding *Crataegus* and other shrubs. *R. arvensis* is southern in distribution, reaching Durham and the Scottish Lowlands but there very rare and sometimes introduced, and occurring throughout much of Ireland.

R. arvensis differs from all other British rose species in its sprawling habit and in the vivid contrast between the glaucous-green and wine-red parts of its stems. Anthocyanin pigmentation is found in other species, but there the glaucous/wine-red contrast is never so pronounced. Even when supported by other shrubs, as in a hedgerow, the long, trailing, young stems with long internodes are apparent and may hang vertically downwards. It is recognised by its long, glandular pedicels, its small, ovoid hips, each with flat disc and fused protruding styles, and by its almost simple sepals. This is the only species in our area with sepals that are not normally green.

R. arvensis 4

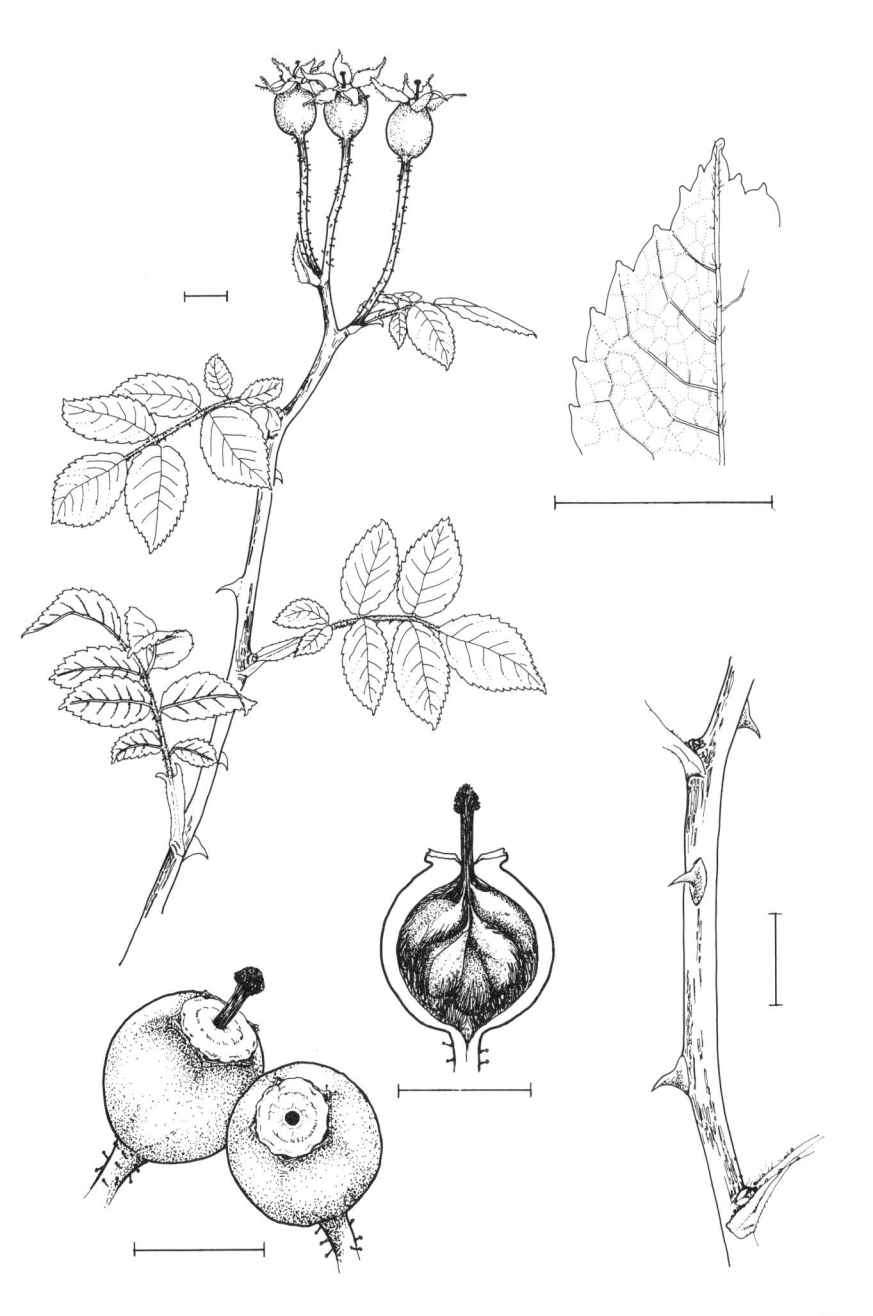

4 × 10. Rosa arvensis Huds. × R. gallica L.
= R. × alba L.

White Rose of York

Erect to spreading shrub to 1.5m; stems green with a pruinose bloom; prickles scattered, slender or stout, of unequal size, slightly curved. Leaflets 5-7, 2.5-6 × 2-5cm, ovate to broadly elliptic or suborbicular, dull green and glabrous above, hairy at least on veins beneath, deeply and sharply serrate; petiole and rachis furrowed, hispid, with scattered glands and short prickles; stipules 25-30 × 4-5mm, large, with spreading auricles. Flowers 1(-3), 6-8cm in diameter, double or semi-double, white to pale pink. Hips 1.7-2 × 1.3-1.7cm, broadly ovoid, red on ripening, some small, brown and aborted; pedicels 3-5cm, thick, strongly glandular-hispid; sepals long, pinnate with expanded leafy tips, glandular-hispid, reflexed and soon falling; stigmas in a villous head, sunk in the orifice or partly concealing the disc; disc slightly concave, orifice 1/3 diameter of disc or more. $2n = 42$. Flowering June – July.

Little-grown today and rarely found in the wild, but recorded as naturalized in S.E. Yorkshire.

V.-c. 61.

4 × 11. Rosa arvensis Huds. × R. stylosa Desv.
= R. × pseudorusticana Crépin ex Rogers   Map 3

General habit of *R. arvensis*, though somewhat stronger and more upright in growth. Leaves are often larger in overall size, the leaflets being more widely spaced and coarsely serrate. Flowers are pale pink, somewhat larger than those of *R. arvensis*. Hips are variable on the same plant: some larger and more elongate than those of *R. arvensis*, others smaller, narrower and sometimes partly sterile. In well-developed hips the disc is slightly, but unmistakeably conical with a somewhat thicker, less protruding stylar column. Pedicels are long and densely glandular. Sepals are often lobed.

V.-c. 3,5,6,9,17,34,36,37.
Non-directional records of the hybrid from v.-c. 13,29,35,44,H21.

R. x **alba 4** x **10**

4 × 12. Rosa arvensis Huds. × R. canina L.
= R. × verticillacantha Mérat   Map 4

General habit of *R. arvensis* but conspicuously more robust and more upright in growth; young stems are wine-red but to a lesser extent. Leaflets are more rounded than in *R. arvensis*, with teeth more acute and less rounded; the leaflets are occasionally biserrate. Flowers are sometimes pink. Hips are frequently sterile and abortive, but when fully developed are larger and more elongated than those of *R. arvensis*, with longer, slightly lobed sepals. Pedicels are variable, 1-3cm, some being almost glandless.

V.-c. 2-7,9,10,13-15,17,19-24,26,29,31-39,41,44,49,55,57,58,60,63,H5, 10,11,19,23,25,28,38,39.
Non-directional records of the hybrid from v.-c. 42,52,73,H21.

4 × 15. Rosa arvensis Huds. × R. tomentosa Sm.

With the long trailing stems and general habit of *R. arvensis*. Distinguished by the stout arcuate or inclined prickles and the narrow lanceolate or ovate-lanceolate acute leaflets, 2-3 × 1-1.5cm, which are thinly pubescent above, tomentose beneath with subfoliar glands, and irregularly glandular-biserrate.

Differs from the hybrid with *R. sherardii* in the larger, coarser leaflets, the stout-based inclined prickles and the extremely small glands.

Only one specimen has been seen by us.

V.-c. 35.

4 × 16. Rosa arvensis Huds. × R. sherardii Davies

Habit and foliage as in *R. arvensis* but distinguished by its deep pink flowers. Leaflets are similar in shape and dentition to those of *R. arvensis* but are slightly bluish and rugose above, and rather sparsely tomentose beneath. Hips are completely sterile and abortive, falling off from the base of the long, glandular-hispid pedicels almost as soon as they have formed.

This appears to be a rare hybrid with only one record.

V.-c. 55.

4 × 18. Rosa arvensis Huds. × R. rubiginosa L.
= R. × consanguinea Gren.

General habit of *R. arvensis*. Leaflets are usually glandular-biserrate and have Sweet-briar-type glands beneath. Such sweet-smelling glands may be found on other parts of the shrub, especially the stipules and sepals. Flowering branches and sometimes pedicels may have patches of small, but strong acicles. Pedicels are somewhat shorter than is usual in *R. arvensis*.

V.-c. 36,38,55,57.
Non-directional records of the hybrid from v.-c. 16.

[4 × 19. Rosa arvensis Huds. × R. micrantha Borrer ex Sm.
= R. × inelegans W.-Dod]

We have seen no specimen, though the reciprocal has been recorded.

5. Rosa pimpinellifolia L. Map 5

Burnet Rose

Low, erect shrub 10-40(-120)cm, suckering and often forming dense thickets rarely more than knee high; prickles numerous, patent or inclined, usually slightly curved, slender but strong with gradually narrowed base, mixed with and passing into acicles. Leaflets (7-)9-11, 0.5-1.5(-2) × 0.5-1.2cm, ovate or suborbicular, obtuse, usually crenate-serrate and eglandular, green, glabrous or slightly pubescent and lighter green beneath; stipules 10 × 4mm, sometimes (mainly in cultivated forms) sparsely glandular (together with petiole and rachis). Flowers solitary, 2-4cm in diameter, creamy white, without bracts. Hips 0.5-1.5(-2) × 0.5-1.2cm, subglobose, purplish-black when ripe, smooth, woody, ripening early; pedicels 1-2(-2.5)cm, smooth or occasionally with gland-tipped acicles; sepals 1.5cm, lanceolate, simple, eglandular, erect when the hips are ripe and persisting until the hips decay; stigmas in a large pilose head, sunk in the orifice; disc slightly concave; orifice large, 1/2 diameter of disc or more. $2n = 28$. Flowering May – July.

Flourishes best on coastal dunes and sandy heaths near the sea. Inland it occurs mainly on chalk and limestone. Found throughout Britain and Ireland though rare in some counties and absent from many inland areas.

This species, with its low erect habit, purplish-black hips, simple erect sepals, dense mixed armature and small neat leaflets cannot be mistaken for any other British species. However it readily hybridizes with other species, and this should be borne in mind if any particular specimen appears to depart from the normal. Garden cultivars are also frequent and may cause confusion. They are often more robust and more glandular than the wild type, and may have pink flowers. Native forms on sand dunes in Ayrshire, the Isle of Arran and in Ireland may also have pink flowers.

R. pimpinellifolia 5

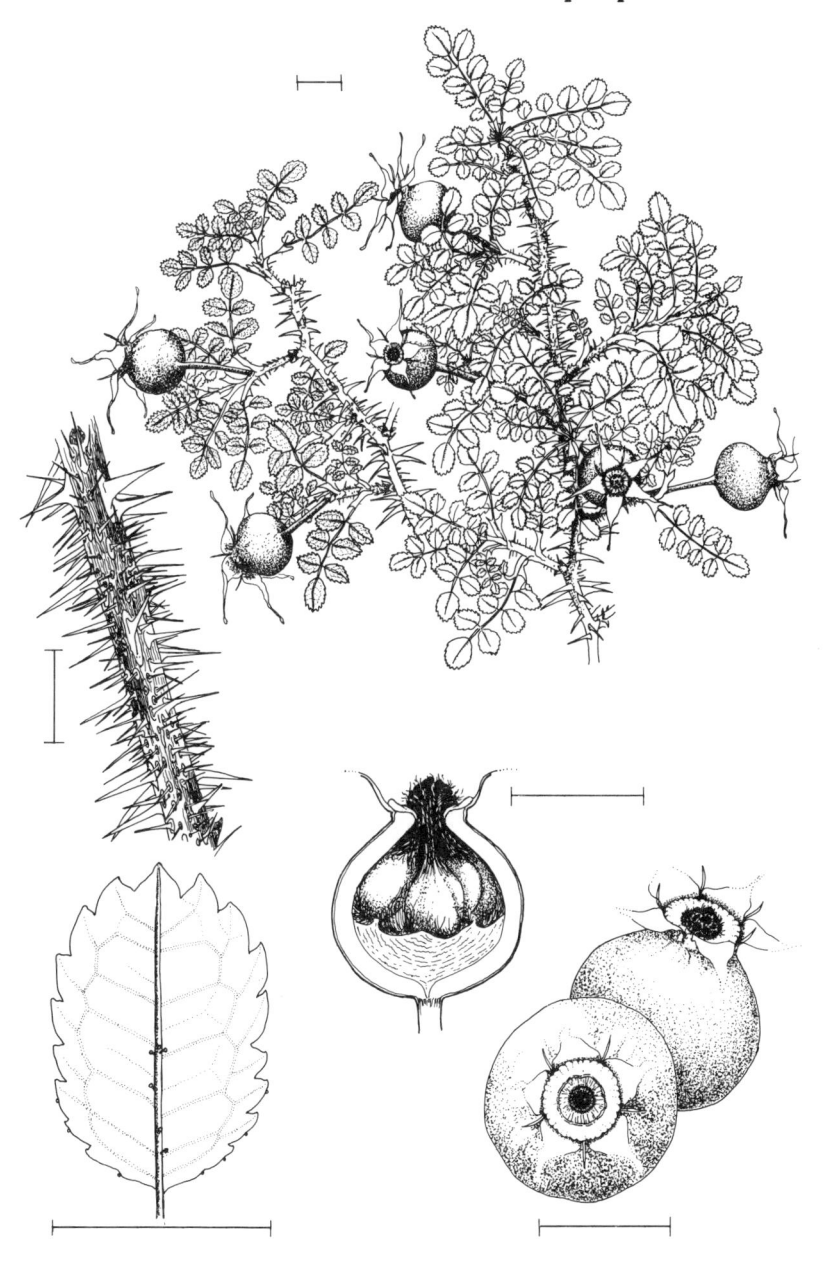

5 × 12. Rosa pimpinellifolia L. × R. canina L.
= R. × hibernica Templeton   Map 6

Habit suckering, with straight or arching stems and some curved prickles mixed with varying quantities of nearly straight, narrow-based prickles as well as acicles. Leaflets, usually 5-9, are glabrous, or sparsely pubescent on the veins beneath, suborbicular, but sharply uniserrate. Stipitate glands may be scattered on leaflet-teeth and stipules. Hips are ovoid, slightly urceolate, or small and subglobose; some may be ill-developed.

In northern England and in Scotland it is often difficult to distinguish this hybrid from the very similar hybrid of *R. pimpinellifolia* with *R. caesia*.

V.-c. 1,17,58,62,64-66,70,82,H38,39.
Non-directional records of the hybrid from v.-c. 16,52.

5 × 13. Rosa pimpinellifolia L. × R. caesia Sm. sensu lato
= R. × margerisonii (W.-Dod) W.-Dod

In densely suckering masses, but differing from *R. pimpinellifolia* in several respects and most particularly in the very mixed armature, strongly curved prickles being present amongst the nearly straight, unequal, narrow-based prickles of the female parent; leaflets are elliptic or obovate and uniserrate or biserrate. The hips have smooth, short pedicels.

It seems pointless to try and distinguish the subspecies of *R. caesia* involved as male parent in this hybrid, though forms with completely glabrous leaves can no doubt be referred to subsp. *glauca*.

The hybrid has been mistakenly recorded, in Scotland and northern England, as *R. × hibernica*, or as a 'variety' of it, but *R. × hibernica* has *R. canina* as one parent. Care should be taken as the distinction is by no means always clear.

V.-c. 66,82,107,H39.
Non-directional records of the hybrid from v.-c. 58,62,108,H38.

5 × 15. **Rosa pimpinellifolia** L. × **R. tomentosa** Sm.
= **R.** × **coronata** Crépin ex Reuter

Habit erect though somewhat arching. Some strong *tomentosa*-type prickles are usually to be found among the characteristic armature of the female parent. Leaflets are larger than in *R. pimpinellifolia*, somewhat elliptical, with more acute teeth, sometimes biserrate, sparsely hairy above, hispid beneath, at least on the veins; petiole and rachis are pubescent, with subsessile glands and occasional pricklets. Hips are small, broadly ovoid or obovoid, and weakly glandular-setose; pedicels are 1.5-3cm, with few or many gland-tipped acicles; sepals are spreading to reflexed, falling early, irregularly pinnate and rather sparsely glandular beneath.

Non-directional records of the hybrid from v.-c. 49,50,H39.

5 × 16. Rosa pimpinellifolia L. × R. sherardii Davies
= R. × involuta Sm.   Map 7

General habit of *R. pimpinellifolia*. Armature is very mixed, slender curved prickles with stouter bases being usually present amongst the narrow-based unequal prickles typical of *R. pimpinellifolia*. Leaflets usually 7(-9), suborbicular to broadly ovate or elliptic, biserrate, hairy and often glandular on the veins beneath. Petiole and rachis are subglabrous to sparsely pubescent with some glands and small arching pricklets. Hips are subglobose, often covered with weak acicles, and blackish. Pedicels are 1-1.5(-2.5)cm and usually densely glandular-aciculate. Sepals are spreading-erect, simple, dorsally glandular, and eventually falling. Flowers are pink.

V.-c. 16,17,38,50,58,62,79,89,90,106,108,H26,39.
Non-directional records of the hybrid from v.-c. 37,42,44,48,49,52,57, 62,64,66-70,82,84,94-96,100,110,H8,12,14,20,21,35,38.

5 × 17. Rosa pimpinellifolia L. × R. mollis Sm.
= R. × sabinii Woods   Map 8

Forming strong, upright, suckering thickets, rather like *R. pimpinellifolia*, but much taller. Stem armature is mixed, though the straight, patent prickles of *R. mollis* do not always stand out amongst the dense, rather similar, but narrow-based prickles of *R. pimpinellifolia*. Wine-red coloration of stems, leaves and bracts is very strong. Leaflets number 7-9, are glandular-biserrate, and sparsely, and softly tomentose; scattered subsessile glands may be hidden amongst the tomentum. Petiole and rachis are pubescent and sparsely glandular. Hips are 1.5 × 1-1.5cm, ovoid-urceolate, often glandular-setose. Pedicels are 1-2cm, sparsely glandular-aciculate. Sepals are erect, often somewhat fleshy at the base, and persistent.

V.-c. 55,66,69,82,88-90,92,94,100,106.
Non-directional records of the hybrid from v.-c. 16,37,57,62,64,65,67, 91,95,96,107-109,111,H9,39.

5 × 18. Rosa pimpinellifolia L. × R. rubiginosa L.
= R. × cantiana (W.-Dod) W.-Dod   Map 9

Habit as in *R. pimpinellifolia*, but with very strong, upright, suckering stems which may reach 2.5m. Armature is very mixed, with large, curved prickles standing out amongst the general clothing of slender, nearly straight prickles and acicles. Leaflets are small, suborbicular to elliptic, with crenate-serrate or glandular-serrate margins, sparsely pubescent and dotted with viscid, Sweet-briar-type glands beneath. Hips are subglobose to ovoid, glandular-setose and with erect, nearly entire, gland-dotted sepals. Pedicels are 0.7-1.5cm.

V.-c. 15-17,80,82,88-90.
Non-directional records of the hybrid from v.-c. 57,59,64,66,67,79,95, 106,H16,26,38.

6. **Rosa rugosa** Thunb. ex Murray   Map 10

Japanese Rose

Erect shrub 1-2m, suckering; stems tomentose when young, stout, densely prickly; prickles subulate, straight, of all sizes down to acicles, the larger prickles pubescent. Leaflets 5-9, 2-5 × 1.5-3cm, large, widely elliptical, bluntly and simply crenate-serrate, edge of teeth often deflexed, dark-green and conspicuously rugose above, pubescent and grey-green beneath; petiole and rachis densely pubescent, with many unequal subulate prickles; stipules 25 × 10-15mm, divergent. Flowers usually solitary, 6-8cm in diameter, bright purplish-pink, white in some cultivars. Hips 1.8-2 × 2-2.5cm, large, depressed globose, red when ripe, smooth; pedicels 1.5-2cm, tomentose, often with gland-tipped acicles, curved in fruit so that the hip points downwards; sepals 2-3cm, long, simple, with a broad, expanded tip, glandular and aciculate, erect on the hip; styles pilose; stigmas in a large domed head, sunk in the concave disc; orifice large, at least 1/2 diameter of disc. $2n = 14$. Flowering June – July.

Commonly grown in gardens as a hedging plant, and often planted as a landscape feature in public places. Occurs occasionally inland as a garden escape and more frequently by the sea on fixed sand-dunes and shingle, where with its suckering habit it may form large thickets. In the map, garden escapes have not been distinguished from deliberate introductions. Native of East Asia.

6 × 12. **Rosa rugosa** Thunb. ex Murray × **R. canina** L. = **R. × praegeri** W.-Dod

Somewhat suckering; stems with straight, slender, patent prickles and a few acicles, along with a few broad-based, curved prickles, the prickles all glabrous. Leaflets (5-)7-9, elliptical, simply and subacutely serrate, edges of teeth often deflexed, dull and slightly rugose on the upper surface and pubescent on the veins beneath; petioles and rachis pubescent or glabrous, usually unarmed; stipules 20 × 10-15mm. Flowers 1-3, 5-7cm in diameter, deep crimson. Hips 1.5-2cm, globose to broadly ovoid; pedicels 2-2.5cm, smooth or with scattered acicles; sepals long, simple or with small lobes and with an expanded tip, glandular-setose beneath; stigmas in a woolly head sunk in the concave orifice; orifice 1/3 diameter of disc.

In the few specimens we have seen of this hybrid *Rosa rugosa* appears to be the female parent.

V.-c. 28,H16,39.

R. rugosa 6

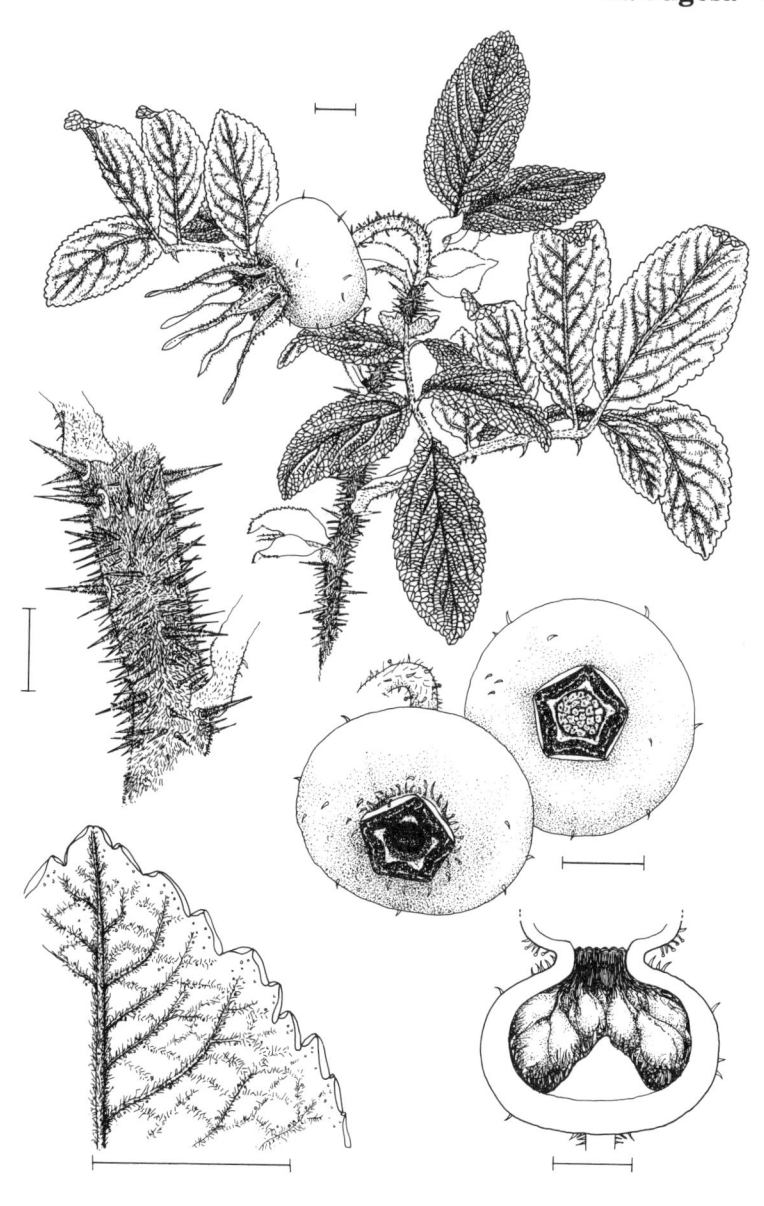

7. Rosa 'Hollandica'

Dutch Rose

Freely suckering and forming dense thickets of uneven height; stems densely covered with yellowish, unequal, slender, declining or slightly curved prickles and acicles, the larger prickles thinly pubescent. Leaflets 7(-9), 3-5.5(-6) × 1.2-2(-2.5)cm, large, elliptical, somewhat narrowed at base, very shallowly crenate-serrate and rugose, bright green and very sparsely hairy above, thinly pubescent beneath; petiole and rachis sparsely glandular and pubescent, stipules 15 × 5-10mm. Flowers solitary, 5-10cm in diameter, dark red. Hips 0.8-1.5cm, globose to ovoid, dark-red when ripe, glandular; pedicels 1.5-2.5cm, densely glandular-hispid; sepals 2-2.5cm, broad at base, tapering to fine tips which are often furcate, densely glandular-hispid, spreading-erect, falling late; stigmas in a large hemispherical, villous head covering the flat disc; orifice 1/3 diameter of disc.

A hybrid of *R. rugosa* from which it differs in having lighter green, smoother leaflets which are less pubescent beneath, and smaller hips.

The plant is of garden origin and as the second parent is not known, its cultivar name is used in the absence of a binomial. Widely used as a rootstock, especially in wet ground, and often naturalized in quantity from garden outcasts on waste ground and in hedges.

V.-c. 6,17,46,99.

R. 'Hollandica' 7

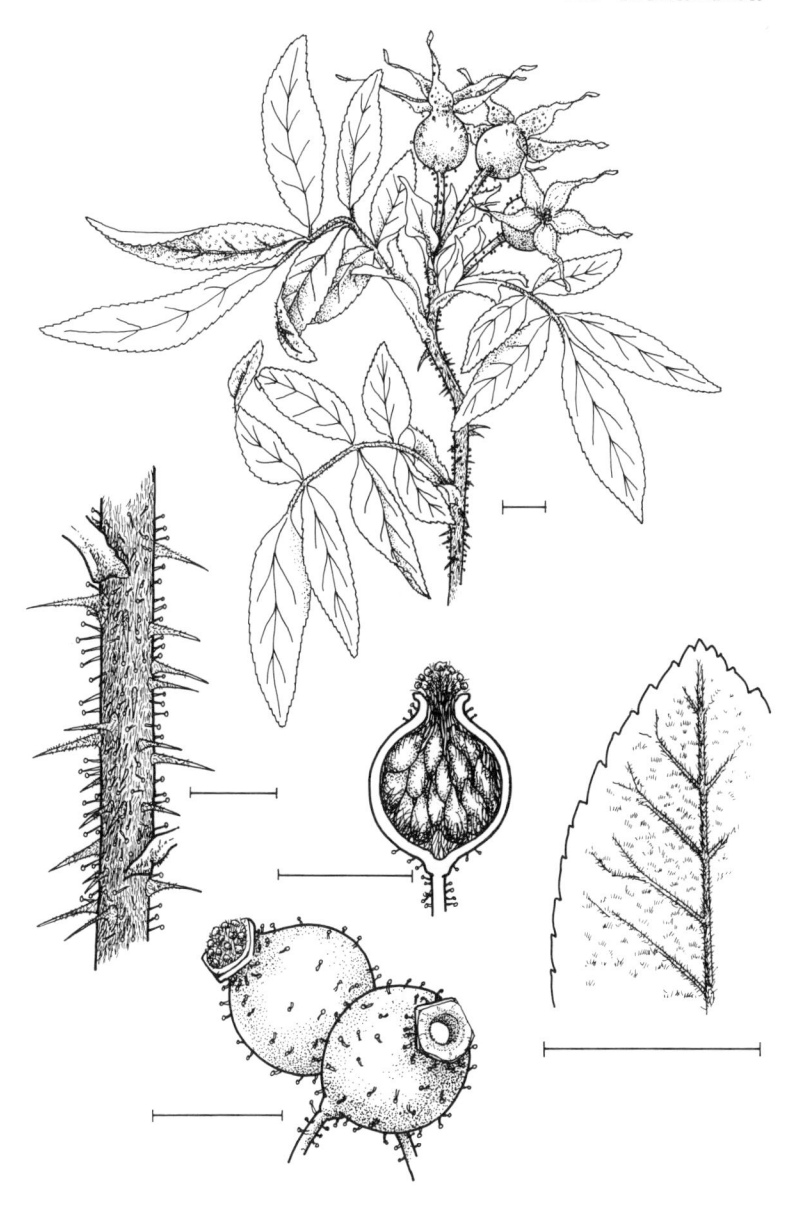

8. Rosa glauca Pourret

Red-leaved Rose

Erect shrub to 3m, suckering; young stems somewhat flexuous, red-brown with a purplish bloom; prickles few, straight or curved; sucker shoots with many acicles. Leaflets 5-7, 2-3.5 × 1-1.5cm, ovate-lanceolate, sharply uniserrate, glabrous, glaucous or suffused with crimson-red; stipules 15 × 2mm, very narrow, sparsely gland-fringed. Flowers few to many, 3-4cm in diameter, clear pink to dark pink. Hips 1-1.5cm, globose, sparsely glandular-hispid, red when ripe; pedicels (1-)1.5-2.5cm, usually glandular hispid; sepals 1-1.5cm, narrow, entire or with a few lateral appendages, spreading-erect, falling after the hips are ripe; stigmas in a villous head; disc 3-4mm, narrow, slightly concave; orifice 1/2 diameter of disc. $2n = 28$. Flowering May-June.

Whole plant conspicuous by its reddish tinge, described variously as brownish-purple and coppery or purple-glaucous. In the leaflet the purple pigmentation often originates in the veins and teeth above, and spreads to the lower surface so that the whole lamina is suffused with crimson.

Naturalized in scattered sites in England and Scotland probably from bird-sown seed. Native of Central Europe.

V.-c. 3,5,29,35,66,95.

R. glauca 8

9. **Rosa virginiana** Herrm.

Virginian Rose

Erect shrub to 1.5m; stems glaucous, sometimes red-brown; suckers few (though the commonly cultivated forms do sucker freely); prickles few, slender, straight or curved. Leaflets 5-7(-9), 2-3 × 1-1.5cm, obovate, narrowed at base, simply serrate, glabrous, dark green and glossy above, lighter green beneath, midrib and veins often tinged with crimson; stipules 20 × 1mm, glossy, narrow, with wider, divergent auricles; petiole and rachis channelled, glossy, tinged with crimson. Flowers 1-5, 4-6cm in diameter, pink or white. Hips 1-1.5cm, subglobose, red or orange on ripening; pedicels 0.8-1.5cm, glandular; sepals narrow, more or less simple, with long, fine tips, densely glandular beneath, spreading-erect, falling as the hips ripen; disc slightly concave; stigmas in a villous or pilose head, partly covering the disc; orifice 1/2 diameter of disc. $2n = 28$. Flowering July-August.

According to Bean (1980) the nomenclature and typification of this plant are in need of verification. Naturalized in scrub and hedgerows in a few sites in England. Native of eastern North America.

V.-c. 5,12,22,59,90.

R. virginiana 9

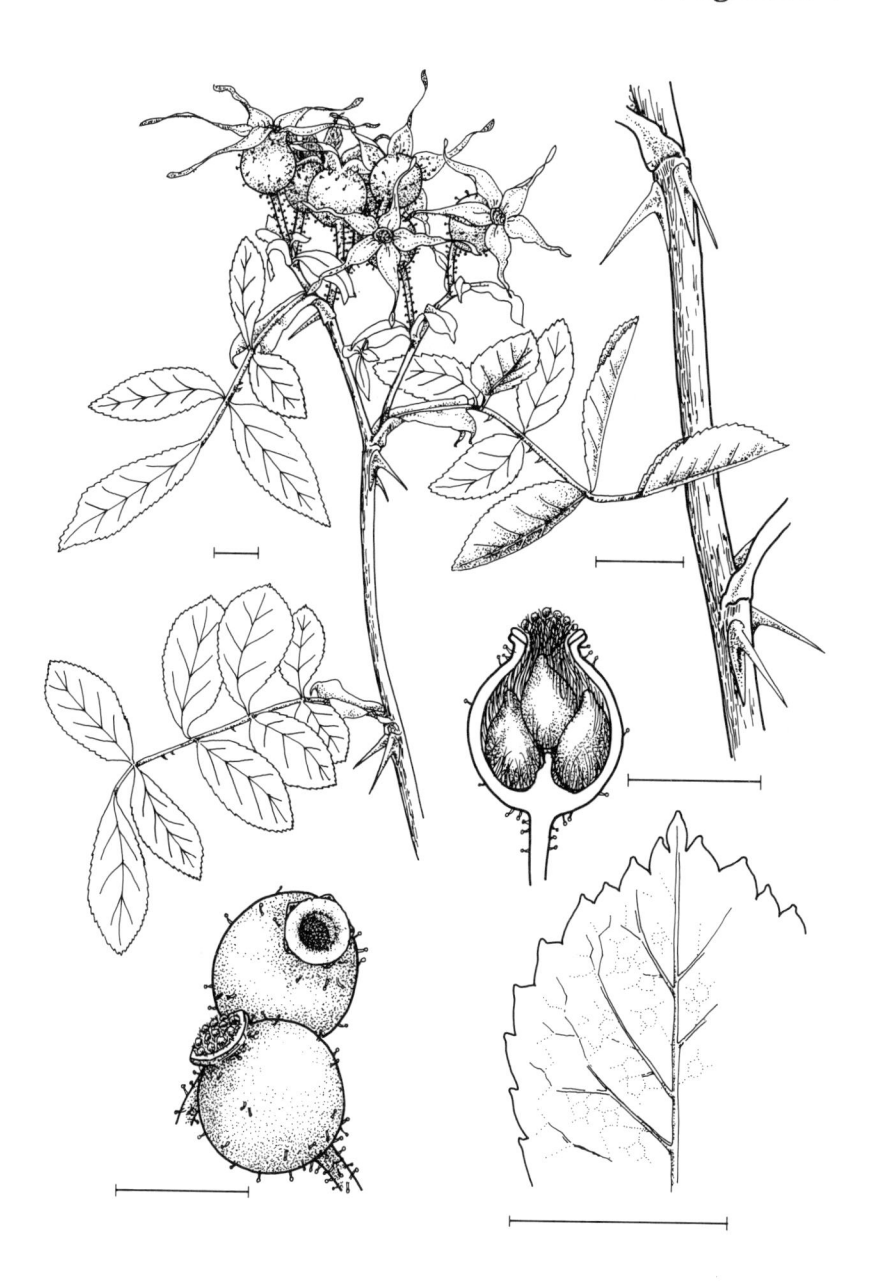

10. Rosa gallica L.

Red Rose of Lancaster

Erect shrub 0.7-1.5m; stems green-glaucous to dull red; prickles many, slightly curved, mixed with dense, short pricklets, glands and gland-tipped acicles. Leaflets 5-7, 3.5-6 × 2-2.5cm, firm, broadly elliptic, dark green and glabrous above, grey-green, hairy and glandular, at least on the main veins, beneath, reticulation rather prominent, shallowly crenate-serrate; petiole and rachis shortly pubescent and glandular; stipules 15-20 × 5-6mm, gland-fringed. Flowers 1(-3), 5-7cm in diameter, pink to red, sweet-scented, petals overlapping, with a white or yellowish claw. Hips 1-2cm, glandular-hispid, ellipsoid to subglobose, slow to ripen, eventually red; pedicels 3-6cm, stout, with sessile or stalked glands and sometimes with a few acicles; sepals c.10 × 5mm, triangular, with many lanceolate-linear pinnae and a long tip, reflexed after flowering and falling before the hips are ripe; stigmas in a pilose or villous head, partly covering the slightly concave disc; orifice 1/3-1/4 diameter of disc. $2n = 28$. Flowering June-July.

The ancestor of many of our garden roses. At one time there were many varieties but few now remain. Naturalized in Guernsey and in a few places in England (Stace 1991). Native of Europe.

V.-c. CI.

R. gallica 10

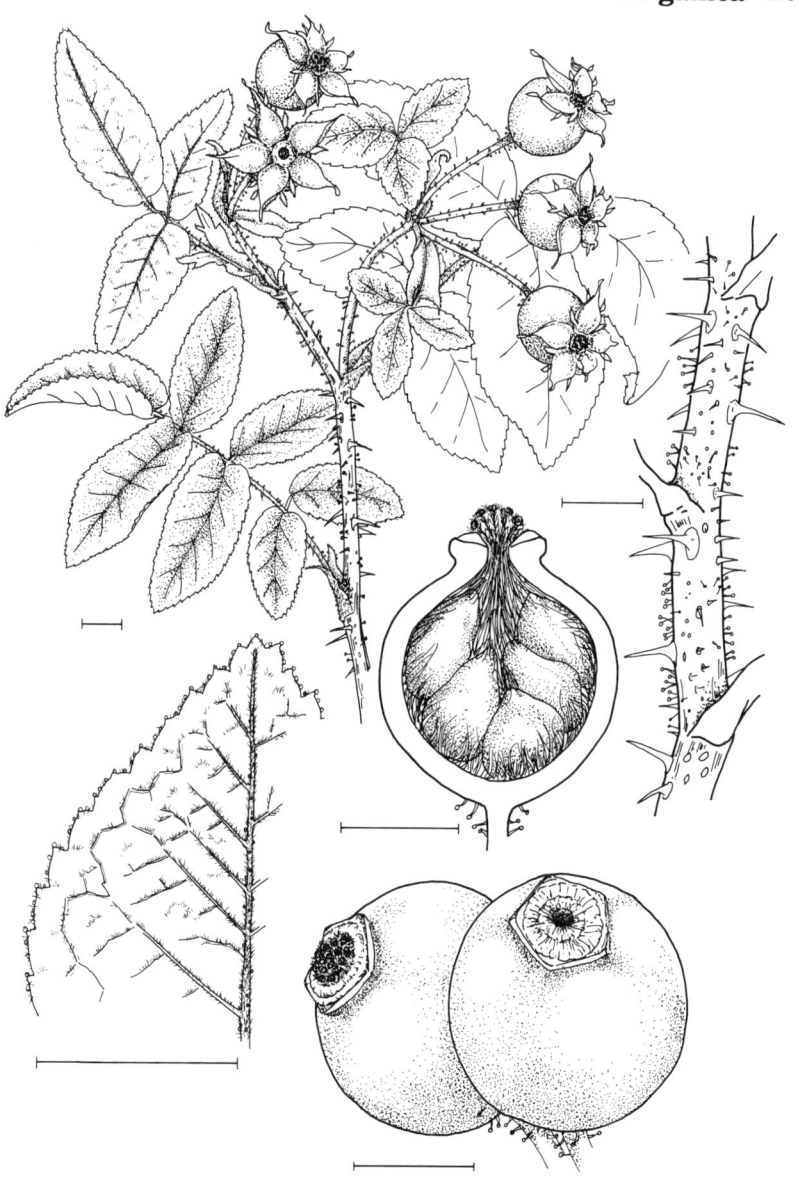

11. Rosa stylosa Desv. Map 11

Columnar-styled Dog-rose, Short-styled Field-rose

Climbing shrub 2-4m, stems arching; prickles deltate, slightly curved above, more strongly curved beneath. Leaflets 5-7, 3-6 × 1.5-2(-3)cm, ovate-lanceolate, tapering to an acute apex, uniserrate, dark green above, light green and pubescent beneath, at least on the veins and occasionally also above (rarely glabrous); petiole and rachis often pubescent, glandular and prickly; stipules 15-25 × 2-5mm, sparsely gland-fringed. Flowers 1-8 or more, 3-5cm in diameter, white or pale pink. Hips 1.2-2 × 0.8-1.5cm, broadly ovoid or globose, red when ripe, smooth; pedicels long, up to 4cm, usually with some stalked glands; sepals 1.5-2cm, sparsely lobed, eglandular, reflexed, falling early; styles glabrous, at first forming a plump, agglutinated, exserted column terminating in a loose, ovoid group of stigmas, later separating and spreading slightly; disc strongly conical; orifice c. 1/5 diameter of disc, continuing with uniform bore down the solid cone of the disc. Flowering June-July. $2n = 35,42$(unbalanced).

Mainly in hedgerows and wood margins. Locally frequent in the south of England and in Wales and in Ireland, rare in the English Midlands and absent from N. England and Scotland.

This species is most easily recognized from a distance by its long, sharply-pointed, dark-green leaflets, which are distantly spaced on the rachis. The deltate prickles on the mature stems are also very characteristic. The thick conical disc of *R. stylosa*, pierced by a hollow cylinder, through which a comparatively long, agglutinated column of styles emerges, is unique amongst British wild roses. It is best seen in section, care being taken to see that the sectioning is absolutely median and vertical. There is, however, a frequent variant of *R. canina* with a conical disc which may cause confusion, especially if the disc is red or orange, as it may be in certain midland counties. However the conical disc in these *R. canina* variants is scarcely thicker than the walls of the hip, the orifice being a mere collar of tissue. In *R. stylosa* the crimson cone of the disc often looks mouldy in autumn as it becomes coated with a thin film of grey tissue.

R. stylosa 11

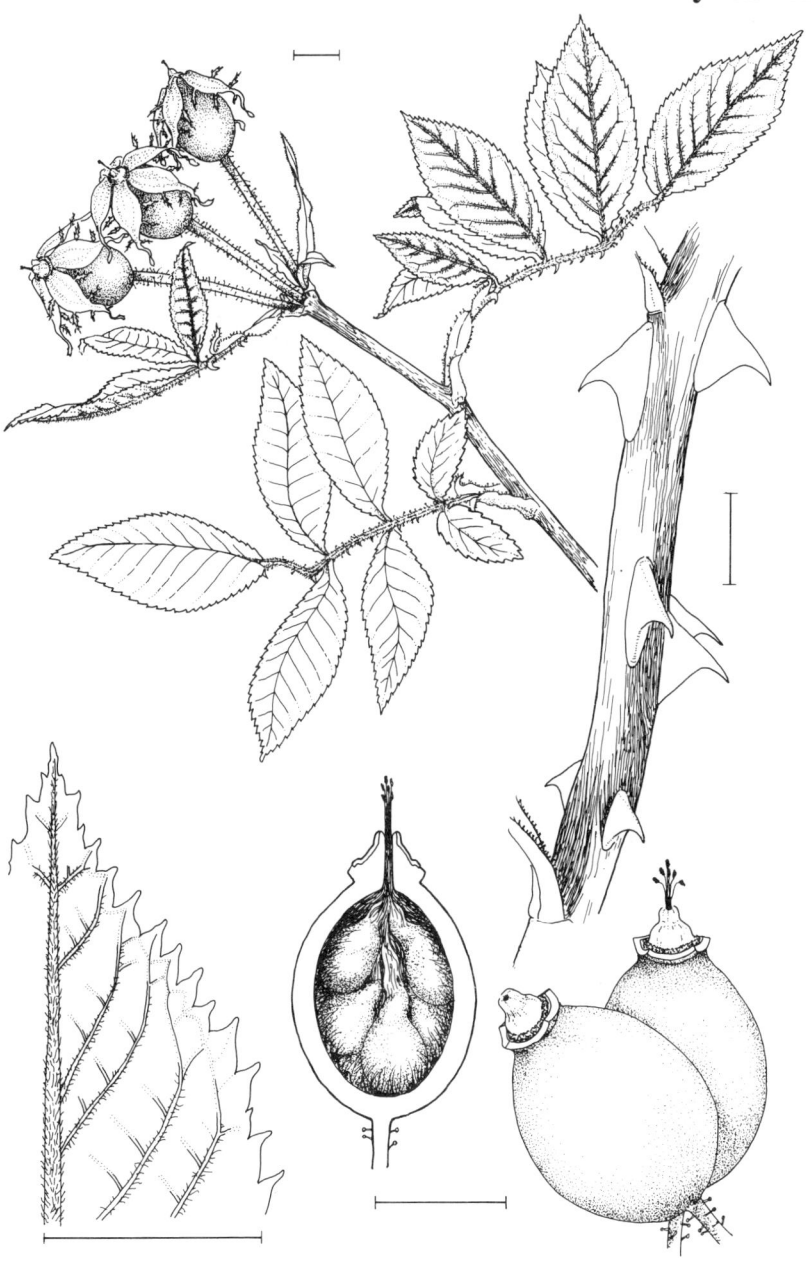

87

11 × 4. Rosa stylosa Desv. × R. arvensis Huds.
= R. × pseudorusticana Crépin ex Rogers Map 3

General habit of *R. stylosa*; though a few strongly pigmented trailing shoots are usually present; prickles sparse, often finely curved. Leaflets coarsely serrate, glabrous or slightly pubescent on the veins beneath. Hips broadly ovoid, medium to small on the same plant; disc conical or convex, styles clumped, or almost fused, not readily separating; long-exserted.

This hybrid, with its strong growth and large, coarsely serrated leaflets is often harder to recognize than its reciprocal, though some strongly pigmented, trailing shoots, so characteristic of *R. arvensis*, are usually present. Hips and styles may also reflect some characters of the male parent.

V.-c. 3,9,37,H21.
Non-directional records of the hybrid from v.-c. 6,13,29,34-36,44.

11 × 12. Rosa stylosa Desv. × R. canina L.
= R. × andegavensis Bast. Map 12

General habit of *R. stylosa*, from which it can only be distinguished with difficulty. Prickles are slightly less stout than those of *R. stylosa*. Leaflets are broader, usually uniserrate and pubescent beneath, occasionally glabrous, biserrate and with stipitate glands beneath. Pedicels are intermediate in length, but may be as little as 1.2cm, sometimes with few or no glands. Styles soon separate and are occasionally hispid.

In parts of southern England and in Ireland where the two species occur together, they readily form hybrids and all intermediates may be found. If it is not obvious which species predominates the name *R. × andegavensis* should be recorded rather than the formula.

V.-c. 1,3,6,9,29,31,34-37,H8,15,18.
Non-directional records of the hybrid from v.-c. 4,5,7,10,11,13,15,17, 21,23,25,27,32,33,38,41,44,46,47,53,61,H2,10,13,14,19,21,39.

11 × 13. Rosa stylosa Desv. × R. caesia Sm.

General habit of *R. stylosa*. Prickles are strongly curved. Leaflets are narrow, glabrous, glaucous beneath, and widely spaced on the rachis. Stipules are short and almost 5mm in width. Hips are as in *R. stylosa*, but with short pedicels only 0.8-1.5cm in length. Sepals are strongly pinnate, short, slightly spreading but falling early. Stigmas are weakly villous and the styles exserted in a fascicle.

The ranges of these species hardly overlap and never, as far as is known, do they occur together in the absence of *R. canina*, so determination of this hybrid is tentative in all cases.

Non-directional records of the hybrid from v.-c. 6,31,34,36,37.

11 × 20. Rosa stylosa Desv. × R. agrestis Savi

A coarse shrub with the general characters of *R. stylosa*. Leaves are variable, some with more or less cuneate bases, many with Sweet-briar-type glands.

Only one bush has been seen by us, the presumed parents being frequent in the locality and *R. micrantha*, which on morphological grounds could have been one of the parents, being absent.

V.-c. 6.

12. Rosa canina L.

Dog-rose

Climbing shrub up to 3 m; young stems green, or sometimes tinged with wine-red, arching; prickles broad-based and strongly curved. Leaflets 5-7, 1-1.8 × 1.5-4cm, ovate or ovate-lanceolate, acute, variously serrate, medium to dark green, glabrous or sparsely pubescent; petiole and rachis usually with a few small prickles, occasionally unarmed; stipules 20 × 6mm. Flowers 1-6, 4-6cm in diameter, pale pink or white. Hips 1-2.5 × 1-1.8cm, very variable, globose, ovoid, obovoid, ellipsoid or urceolate, red when ripe, smooth; pedicels 1.5-2.5cm, smooth; sepals 2-4cm, pinnate, sometimes rising to the horizontal but eventually reflexed, falling early before the hips are ripe; styles glabrous, hispid or slightly pubescent; stigmas in a small globose or conical head; disc flat or convex, occasionally conical; orifice small, 1/5-1/6 diameter of disc. $2n = 35$(unbalanced). Flowering June-July.

Hedgerows and scrub. Often a characteristic early colonizer of the verges and ballast of dismantled railways, quarry spoil-heaps and other open habitats. The commonest rose throughout most of Britain and Ireland, though in Scotland it tends to be replaced by *R. caesia*.

Distinguished from the Downy-roses by its smooth pedicels and from the Sweet-briars by the absence of brown, viscid (normally odorous) glands. It can be distinguished from *R. caesia* by its small, non-villous head of stigmas never concealing the disc, a much smaller orifice and reflexed sepals. Specimens with a somewhat conical disc are distinguished from *R. stylosa* by the leaflets, which are not widely spaced on the rachis, by the absence of glands on the pedicels and by the very different section of the disc (see morphology chapter). Isolated bushes in open habitats grow erect when young, but when more mature form a wide dome-shaped bush with stems arching out from the centre.

R. canina 12

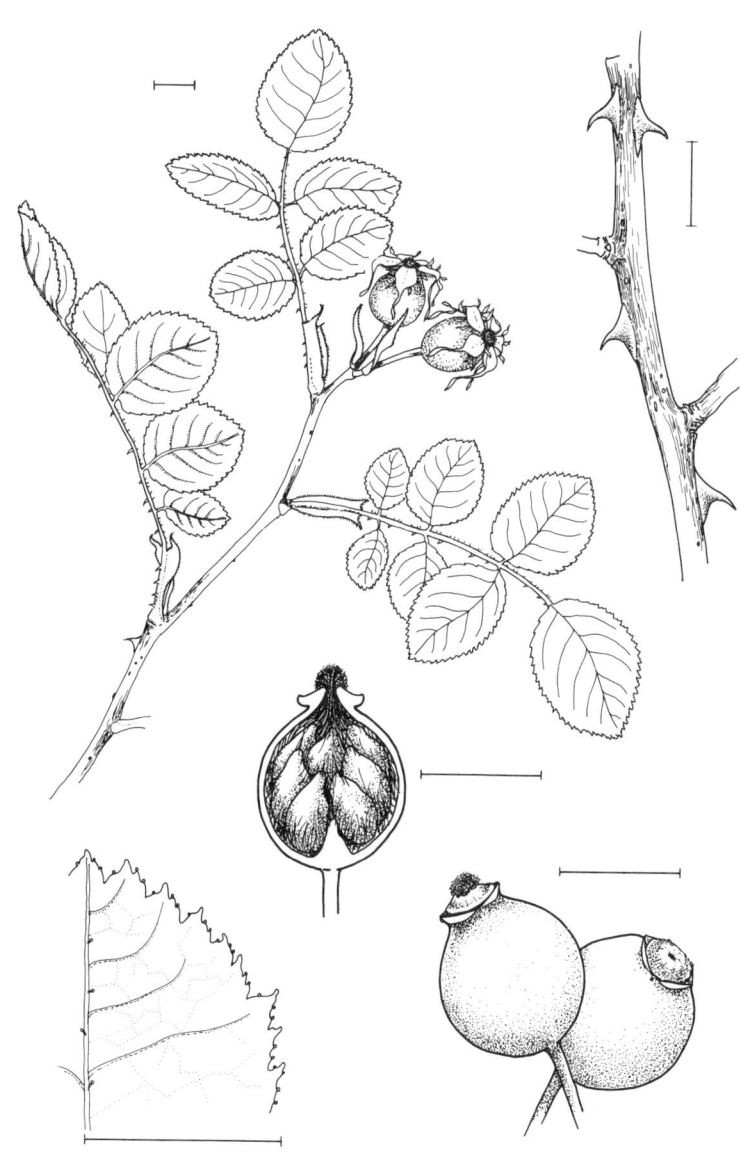

Rosa canina as we understand it is at present very variable. Although there is an almost continuous range of variation, three distinct, extreme types can be recognized. These have been named by other authorities as distinct species, and other treatments will undoubtedly be proposed for them in the future. To allow, therefore, for possible changes in taxonomic opinion, we have retained four of Wolley-Dod's informal Group names for the three extremes and one transitional Group. This should ensure that records made on these lines will be able to be correlated with any reassessment of the species resulting from future research. The majority of specimens can be placed in one or other of these Groups, but as the variation is almost continuous some will be found which are difficult to place. In the regions where *R. canina* occurs there are usually plenty of specimens which can be confidently placed in one of the Groups, and the few which are difficult to place can be ignored for recording purposes.

The distinguishing characters of the four Groups are given below. There is as yet no evidence that any one Group has a different ecological niche or geographical distribution.

Group **Lutetianae** (Corresponds to the lectotype of *R. canina* in LINN) Map 13

Whole plant glabrous (except possibly some hispidity of the styles); leaflets uniserrate and eglandular; stipules, petiole and rachis usually eglandular, or with a few stipitate glands on the margins of the stipules.

Group **Dumales** (*R. canina* L. var. *squarrosa* Rau; *R. squarrosa* (Rau) Boreau) Map 14

Whole plant glabrous (except possibly some hispidity of the styles); leaflets biserrate or multiserrate, with small, scentless, shining red glands on the denticles; stipules densely fringed with small, shining, red glands; petiole and rachis with a few stipitate red glands; occasionally a few similar glands on the lower surface of the leaflets, especially on the midribs.

Group Transitoriae   Map 15

Whole plant glabrous (except possibly some hispidity of the styles); leaflets irregularly uniserrate, that is with large teeth alternating irregularly with small ones, the latter being tipped with dull, dark brown glands or gland rudiments (these smaller teeth are between the larger ones and not on their sides as with a truly biserrate leaflet); stipules with some stipitate glands on the margins; sometimes a few glands on petiole and rachis.

Probably consists of hybrids of the above two Groups, with Group *Lutetianae* as the female parent. The reciprocal hybrid is probably indistinguishable from Group Dumales, though it may include the less typical forms with somewhat irregular biserration.

Group Pubescentes (*R. dumetorum* auct., non Thuill.; *R. corymbifera* Borkh.)   Map 16

Leaflets usually uniserrate and eglandular, pubescent beneath, often sparsely so or with hairs confined to the midrib; petiole and rachis pubescent, sometimes densely so; stipules usually eglandular, or with a few stipitate glands on the margins.

12 × 4. Rosa canina L. × R. arvensis Huds.
= R. × verticillacantha Mérat   Map 4

Barely distinguishable from *R. canina* without close inspection except for the strong development of anthocyanin in some young stems. Prickles are as in *R. canina* but with a tendency to occur in clusters. Leaflets occasionally show some sign of the curved (crenate-serrate) teeth of *R. arvensis*. Hips are occasionally small and narrow, with a flattish disc and a tendency for the styles to fuse. Pedicels are variable in length on the same bush, usually sparsely glandular-hispid.

V.-c. 3,13,17,24,26,30-32,34,36-38,40,46,55,57,58,63,70.
Non-directional records of the hybrid from v.-c. 15,41,42,44,49,52,73, H10,19,21.

12 × 5. Rosa canina L. × R. pimpinellifolia L.
= R. × hibernica Templeton   Map 6

Clearly much nearer to *R. canina* in leaf characters and habit, but the mixed armature, characteristic of the hybrid, is usually present to some extent. A suckering habit may be a good field indicator but we have only seen two herbarium specimens. Although *R. pimpinellifolia* has usually been recorded as the female parent, the reciprocal hybrid does occur. Northern forms recorded as *R.* × *hibernica* are often errors for *R. pimpinellifolia* × *R. caesia*.

V.-c. 62,H39.
Non-directional records of the hybrid from v.-c. 16,17,52,58,65, H38.

12 × 6. Rosa canina L. × R. rugosa Thunb. ex Murray
= R. × praegeri W.-Dod

Only the reciprocal hybrid seems to have been recognized.

12 × 11. Rosa canina L. × R. stylosa Desv.
= R. × andegavensis Bast.   Map 12

General appearance of *R. canina* accompanied by various diluted characters of *R. stylosa*. Leaflets are slightly narrower than in *R. canina*, uniserrate or irregularly biserrate, sparsely pubescent beneath or frequently glabrous, the pairs being spaced a little more widely than those of *R. canina*. Pedicels are glandular-hispid to some extent, variable in length, though some may reach 2-2.5cm. The disc may be slightly thickened and conical.

In parts of southern England and in Ireland where the two species occur together, they readily form hybrids and all intermediates may be found. However it is usually obvious which species predominates. If this is not so then just the binomial *R. × andegavensis* should be recorded.

V.-c. 1-4,6,9,34-37,57.
Non-directional records of the hybrid from v.-c. 5,7,10,11,13,15,17,21, 23,25,27,29,32,33,38,41,44,46,47,53,61,H2,10,13-15,18,19,21,39.

12 × 13. Rosa canina L. × R. caesia Sm.
= R. × dumalis Bechst.   Maps 17 & 18

As noted under the reciprocal hybrid these two species hybridize freely and frequently whenever they are found together. In general it is better to record hybrids under the binomial, *R. × dumalis*, as it is very difficult to elucidate the female parent, and often also which subspecies of *R. caesia* is involved in the cross. Plants near to *R. canina*, but with wine-red pigmentation in stem and leaves, glabrous, uniserrate, glaucous leaflets, large hips ripening early, pedicels rather short, and very hispid stigmas, can be considered to be a hybrid with *R. canina* as the female parent and *R. caesia* subsp. *glauca* as male parent.

Similar plants with less glaucous, darker green leaflets, which are hairy at least on the veins beneath may be considered hybrids with *R. caesia* subsp. *caesia* as male parent. However, unless subsp. *caesia* is common in the area and hairy forms of *R. canina* scarce, such determinations are somewhat conjectural.

V.-c. 35,44,46,89,102.
Non-directional records of the hybrid from v.-c. 3-6,9,11,12,14,16,17-21,23,24,26-29,31-34,36-45,47-53,55,57-62,64,66-70,72-74,80,82-84, 86,88,90,92,95,96,99,100,106,107,111, H5,8,9,11,13,14,19-23,25,31-33, 38-40.

12 × 14. Rosa canina L. × R. obtusifolia Desv.
= R. × dumetorum Thuill.   Map 19

Where these two species occur together there is often a bewildering series of hybrids showing all gradations between the two species. It is therefore virtually impossible to give satisfactory descriptions which would delimit the reciprocal hybrids. The influence of *R. obtusifolia* is usually easy to observe, but the assumption of *R. canina* influence is often a matter of probability, based on knowledge of the roses of a particular region. When the *R. canina* influence is only slight, e.g. if the only departure from *R. obtusifolia* consists of slightly larger flowers with pale pink petals, the specimen can be recorded as *R. obtusifolia*, with a small degree of permissible introgression. *R. canina* with suspected traces of *R. obtusifolia* influence is best ignored for recording purposes, as it is too easy to confuse it with *R. canina* Group *Pubescentes*. Where there is strong evidence of the presence of both species the specimen should be labelled *R. × dumetorum*.

V.-c. 27,29,55.
Non-directional records of the hybrid from v.-c. 4,6,12,13,15,17, 18,21-23,25,26,31-38,40-43,49,50,57,58,60-64,H18,19,21,26,40.

12 × 15. Rosa canina L. × R. tomentosa Sm.
= R. × scabriuscula Sm.   Map 20

Habit of *R. canina*. Leaflets sometimes dull, dark green, long and narrow, sparsely pubescent beneath with some tendency to glandular-biserration. Hips are small, globose, with moderately long, partially glandular-hispid pedicels (some to 2 or 2.5cm). Stigmas are glabrous or sparsely hispid in a very small head. Scarcely to be separated from *R. canina × R. sherardii* except for the long pedicels.

V.-c. 3-5,7,13,16,17,19,20,23,31,32,34,40,43,49,55,57,58,63,69,70,H2, 19,21.
Non-directional records of the hybrid from v.-c. 6,8,11,12,15,22,26,28, 29,35-38,41,42,44,48,53,65,H2,4,5,10,11,14,16,17,19,20,33,40.

12 × 16. Rosa canina L. × R. sherardii Davies
= **R. × rothschildii** Druce Map 21

Habit and armature of *R. canina*. Leaflets are glabrous or slightly pubescent, often suffused with crimson, sometimes glandular-biserrate. Reddish, aromatic *sherardii*-type glands are usually found on the undersurface of the leaflets. At least a few glandular-hispid pedicels are normally present. Sepals are usually reflexed but may fall late. Stigmas are usually very hispid and in a small head.

V.-c. 32,36,43,47,57,58,H5,20,21,28.
Non-directional records of the hybrid from v.-c. 35,37,42,44-46,48-52,62,66,69,70,82,85,94,96,H6,8,12,14,18,19,22,23,29-32,35,36,38,39.

12 × 17. Rosa canina L. × R. mollis Sm.
= **R. × molletorum** H.-Harr.

Habit of *R. canina* and with a mixed armature of curved and completely straight, patent prickles. The leaflets may be completely glabrous or slightly hispid. Sparse resinous glands are often present on leaflets and stipules and sometimes on the sepals. Several of the pedicels may also occasionally have a few glands. The styles and stigmas are usually very hispid.

This hybrid was described from Co. Durham by Heslop-Harrison (1955), but we have been unable to trace the material. It is difficult to separate from *R. caesia* × *R. mollis* in the North. In Scotland the latter is the more prevalent hybrid though both are rare.

V.-c. 66.
Non-directional records of the hybrid from v.-c. 60,67,68,104.

12 × 18. Rosa canina L. × R. rubiginosa L.
= R. × nitidula Besser

General habit of *R. canina*. The armature is very mixed, with coarse, blunt prickles standing out on the main stems against the strongly curved, acuminate prickles typical of *R. rubiginosa*. Acicles are occasionally found on the flowering branches. Leaflets may be glabrous or hispid, often intermediate in shape, with viscous, Sweet-briar-type glands beneath. Stigmas are usually very hispid and in a large head.

V.-c. 17,32,42,55.

Non-directional records of the hybrid from v.-c. 16,28,29,31,38,41,57, 64,66,H39.

13. Rosa caesia Sm.

Northern Dog-rose

Climbing shrub with arching stems reaching 2m; young stems often with wine-red pigmentation on sunlit side; prickles broad-based and strongly curved. Leaflets 5-7, 2-3.5(-4) × (1-)1.5-2.5cm, ovate or elliptical, uniserrate or biserrate, dark green above, caesious or glaucous beneath, pubescent or completely glabrous, occasionally with a few glands; stipules large, 15-20 × 8mm, with short, acute auricles. Flowers 2-4(or more), 3-5cm in diameter, usually pink. Hips 2-3 × 1-2cm, globose, ovoid, obovoid, or ellipsoid, red when ripe, smooth; pedicels 0.5-1(-1.5)cm, very short, smooth, partly hidden by broad, leafy bracts; sepals 1.5-2cm, pinnate, spreading-erect after anthesis but falling when the hips ripen; styles villous; stigmas in a densely villous dome-shaped head which almost completely covers the disc; orifice 1/3 diameter of disc, slightly less in southern forms. Flowering June-July. $2n = 35$(unbalanced).

Found in hedgerows and on road verges, tracksides and hill slopes. Northern in distribution, very common from Yorkshire and Lancashire northwards, decreasing in frequency southwards but reaching the Midlands, occasional in Northern Ireland.

R. caesia can be distinguished from the Downy roses by the absence of glands on pedicels and hips, and by the leaflets which are either glabrous, or if hairy, never tomentose. It can be distinguished from *R. canina* by its larger hips with spreading-erect sepals, short pedicels concealed by wide bracts, a large villous head of stigmas and a larger stylar orifice. Another distinguishing character is the strong wine-red colouration, usually found in both subspecies of *R. caesia* and absent from, or weak in, *R. canina* forms. Indeed this colouration, when found in a bush having many characters of the latter species, is often a good indication of hybridity with *R. caesia*.

R. caesia subsp. caesia 13a

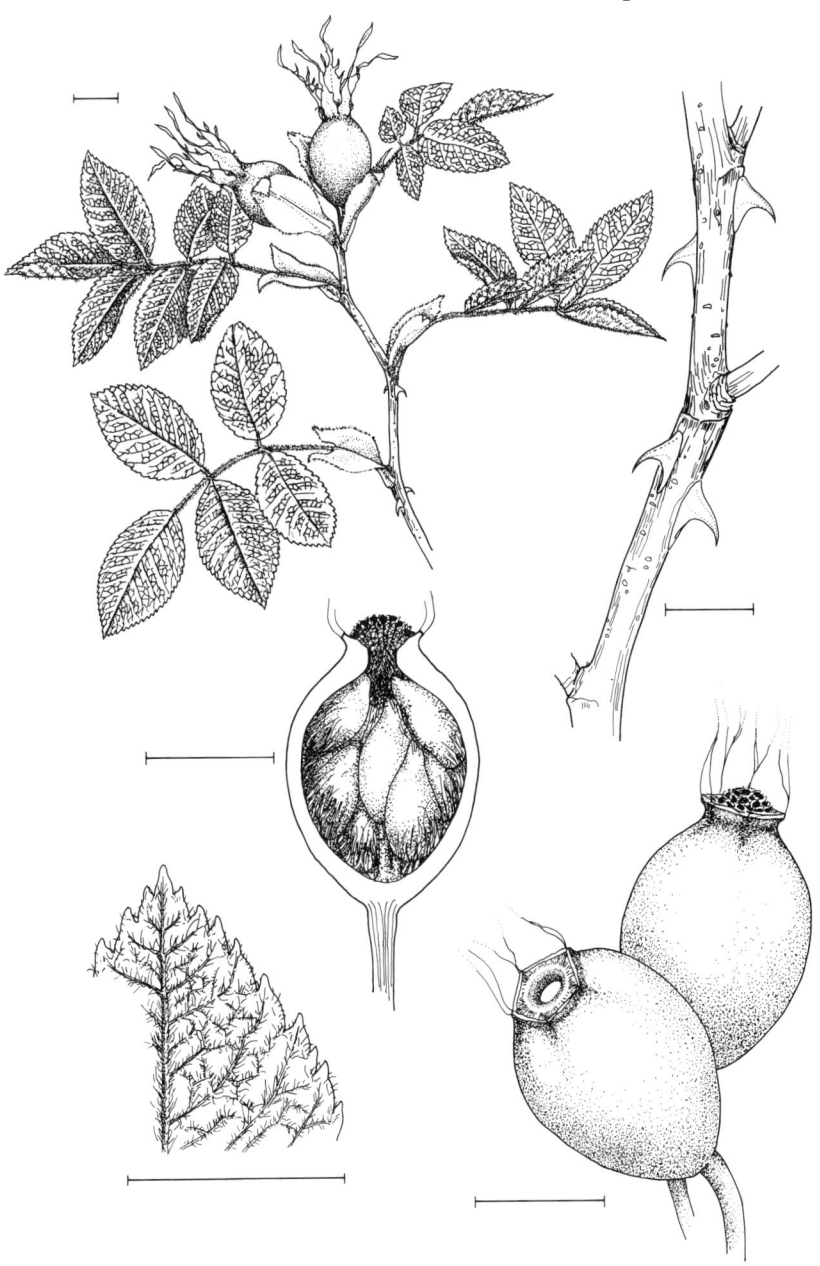

13a. Rosa caesia subsp. caesia   Map 22

Hairy Northern Dog-rose

Leaflets usually ovate, dark green and somewhat rugose above, pubescent at least on the veins and caesious beneath.

Found in hedgerows and on woodland margins, road verges and tracksides. Distribution as for the species. In northern England and the Scottish Lowlands subsp. *caesia* and its hybrids are almost as frequent as *R. canina*.

Differs from subsp. *glauca* mainly in the hairy leaflets, but also in having, when mature, a stronger, more compact growth form, supporting larger clusters of flowers or hips. In southern forms the wine-red pigmentation is often rather weak. In all forms of this subspecies the profusion of caesious leaves tends to mask the wine-red pigmentation of the stems; in subsp. *glauca*, on the other hand, the glaucous undersides of the narrow, often folded leaves stand out vividly against the red stems.

13b. Rosa caesia subsp. glauca (Nyman) G.G. Graham & Primavesi   Map 23

Glaucous Northern Dog-rose

Leaflets narrowly ovate, smooth, completely glabrous, and glaucous beneath.

Found in hedgerows and scrub and on road verges, tracksides and hill slopes. Distribution as for the species, but slightly more northerly and certainly more montane, completely displacing *R. canina* in the Scottish hills.

R. caesia subsp. **glauca** 13b

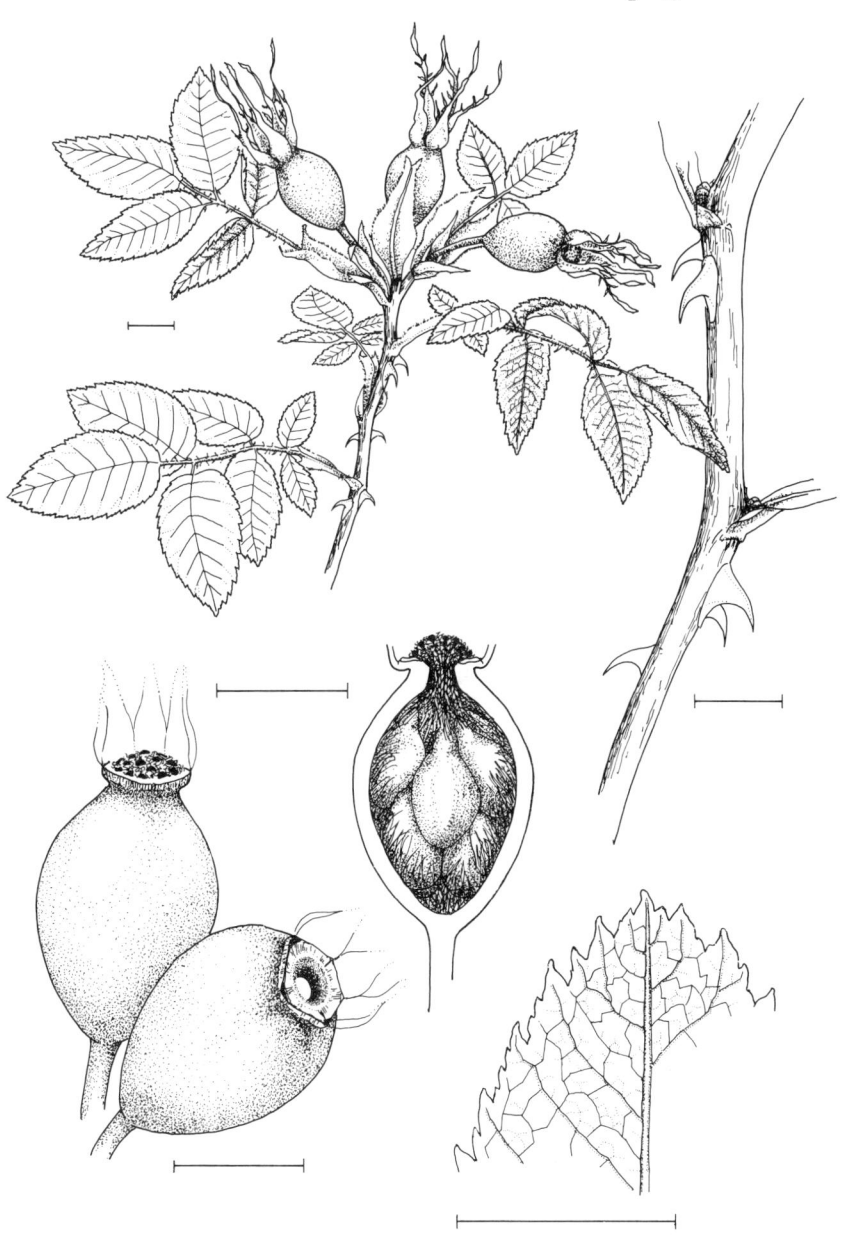

13b × 4. Rosa caesia Sm. subsp. glauca × R. arvensis Huds.

Differs from *R. caesia* in the long, weak leading shoots with long internodes which have more strongly contrasting green shaded sides and wine-red sunlit sides, a feature indicative of *R. arvensis*. Prickles are more slender than is usual in *R. caesia*, with a tendency to clustering. Leaflets are somewhat glaucous beneath, with many of the teeth showing the characteristic curvature of *R. arvensis*. Hips resemble those of *R. caesia*, with a comparatively wide orifice; pedicels are short and glandular-hispid; sepals are short, with small lobes, falling early from the suberect position. Styles are not exserted and the large domed head of stigmas is only slightly villous.

V.-c. 40,55,57.

[13a × 5. Rosa caesia Sm. subsp. caesia × R. pimpinellifolia]

Has not been reliably recorded. Indeed it would probably differ from the previous hybrid only in having slightly pubescent leaflets.

Non-directional records of the hybrid from v.-c. 82.

13b × 5. Rosa caesia Sm. subsp. glauca × R. pimpinellifolia L.

Differs from *R. caesia* subsp. *glauca* in the habit, which is usually more compact and erect, and in the very mixed armature of both main stems and branches. Wine-red pigmentation is fairly strong in all parts of the bush, though the main stems may be blackish. Leaflets are quite glabrous and often glaucous beneath. Flowers are pinkish. Hips are usually aborted or sterile, though a few large ellipsoid hips, crowned with irregularly pinnate sepals, may be present.

V.-c. 66,H39.
Non-directional records of the hybrid from v.-c. 58,62,107,108.

13 × 12. Rosa caesia Sm. × R. canina L. = R. × dumalis Bechst.   Maps 17 & 18

Differs from *R. caesia* mainly in the reflexed sepals which fall early, a somewhat smaller, less villous head of stigmas, a smaller stylar orifice, 1/5-1/4 diameter of disc, and longer pedicels which are not concealed by the bracts; wine-red pigmentation is sometimes much reduced.

This is a common hybrid. In Northern England and in the Midlands it is considerably more frequent than the pure species, and it is sometimes impossible to distinguish the reciprocal hybrids from one another.

The subspecies involved in such hybrids can sometimes be determined by a competent rhodologist who has a thorough knowledge of the area from which the specimens come (see notes under the reciprocal hybrids). If there is any doubt it is better to record the plant as *R.* × *dumalis*.

V.-c. 35,66,90.
Non-directional records of the hybrid from v.-c. 3-6,9,11,12,14,16,17-21,23,24,26-29,31-34,36-53,55,57-62,64,67-70,72-74,80,82-84,86,88,89,92,95,96,99,100,106,107,111,H5,8,9,11,13,14,19-23,25,31-33,38-40.

13b × 15. Rosa caesia Sm. subsp. glauca × R. tomentosa Sm. = R. × rogersii W.-Dod

Habit intermediate between the two species, tall and arching with strongly wine-red stems. Prickles on older stems are stout, broad-based and strongly curved; those on the younger stems are slender and arcuate. Leaflets are long and narrow, greyish-green, not glaucous, weakly biserrate, with some glands on the denticles and sparsely tomentose beneath. Hips are smooth above, glandular-stipitate at the base; pedicels are glandular-hispid; sepals are reflexed and fall early; the stigmas are in a broad villous head. Most of these characters could also indicate a hybrid with *R. sherardii*, but the strong climbing and arching habit, the reflexed sepals falling early, and the stylar orifice smaller than is usual in *R. caesia* point to *R. tomentosa* as the other parent.

V.-c. 38,55,57.

13 × 16. Rosa caesia Sm. × R. sherardii Davies   Map 24

Habit of *R. caesia* with strong stems, the youngest strongly tinged with a brownish, wine-red colouration; prickles are strongly curved, with narrow bases, mixed (on the young branches) with others more slender and arcuate-acuminate. Leaflets are caesious, rugose and glabrous above, rather ashy-grey pubescent or glabrous beneath and with a few small, reddish resinous glands on the veins more or less hidden by any pubescence, glandular-multiserrate; stipules are 15 × 10mm, short and very wide, gland-fringed. Hips are large, obovoid to subglobose; pedicels 3-7mm, very short, sparsely glandular-hispid or glabrous; sepals rising to the horizontal or a little higher on the young hips, dorsally glandular, falling late; stylar orifice c. 1/3 diameter of disc.

It is sometimes possible, though difficult, to separate the subspecies of *R. caesia* involved in the cross.

V.-c. 49,55,62,66,88,106,107,111.
Non-directional records of the hybrid from v.-c. 42,52,59,69,70,82,83, 92,95,96,98-100,103,105,H22,29,32,33,35,38-40.

13 × 17. Rosa caesia Sm. × R. mollis Sm. = R. × glaucoides W.-Dod

Habit and armature of *R. caesia* but with occasional slender, straight, patent prickles. The main stems are strongly tinged with wine-red. Leaflets are often broadly elliptical, slightly hispid and biserrate. Resinous glands may be scattered on the leaflets, rachis, stipules and sepals. Hips are large, occasionally with erect sepals; pedicels are very short, occasionally glandular-hispid, hidden by large, leafy bracts; orifice is large; stigmas in a large, villous, domed head.

V.-c. 57,105.
Non-directional records of the hybrid from v.-c. 66-68,70,95,96,111.

13 × 18. Rosa caesia Sm. × R. rubiginosa L.

Habit intermediate, with very strong, reddish stems, at first upright, but finally arching; prickles on main stems stout and curved but mixed with some more slender and arcuate on the flowering branchlets. Leaflets are mixed, some obovate and in serration near to those of *R. rubiginosa*, some elliptic and more sharply serrate; all are glabrous above, and pubescent as well as viscid-glandular beneath. Hips are large, subglobose to obovoid with spreading sepals which fall late in the season. The description refers to the hybrid with subsp. *caesia*.

Plants as above, but with narrowly ovate leaves, more or less glabrous except for a few Sweet-briar-type glands, may be presumed to have *R. caesia* subsp. *glauca* as female parent.

V.-c. 67,89.
Non-directional records of the hybrid from v.-c. 57,66,68,90,94-96.

13 × 19. Rosa caesia Sm. × R. micrantha Borrer ex Sm. = R. × longicolla Ravaud ex Rouy

Habit and armature of *R. caesia*. Leaflets are smaller and narrower with varying quantities of Sweet-briar-type glands beneath. Hips are much smaller than is normal in *R. caesia*, with a small head of hispid stigmas. Sepals are reflexed and falling early. Orifice is c. 1/4 diameter of disc.

The absence of acicles on any part of the bush as well as the absence of any strongly arched prickles, coupled with the presence of narrow leaves and early-falling sepals rules out the possibility that *R. rubiginosa* is the male parent.

V.-c. 42.

14. Rosa obtusifolia Desv.   Map 25

Round-leaved Dog-rose

Usually low-growing shrub 1-2m, with arching stems; prickles broad-based, strongly curved, abruptly contracted to a long fine point. Leaflets 5-7, 2-3(-3.5) × 1.5-2cm, broadly ovate with rounded base, subacute, dark green, pubescent at least beneath but not tomentose, often glandular on lower surface, finely biserrate with numerous small, reddish-brown glands on the teeth (all glands odourless); petiole and rachis pubescent and often glandular; stipules 15 × 4mm, with glandular margins and narrow, acute auricles. Flowers 2-5, 3-4cm in diameter, white. Hips 1-1.5 × 0.8-1cm, broadly ovoid or globose, red when ripe, smooth; pedicels 0.5-1.5cm, eglandular; sepals 1.5-2cm, bipinnate with large lobes, strongly reflexed after flowering and often concealing the hips, falling early; disc convex; stigmas in a small, globose head; styles and stigmas hispid or subglabrous; orifice small, 1/6-1/5 diameter of disc. $2n = 35$(unbalanced). Flowering June-July.

Hedgerows and scrub, frequent in the southern part of England, becoming rarer northwards from the Midlands and absent from Scotland. Rare in Wales and in Ireland.

R. obtusifolia is most easily separated by the rather neat-looking, often overlapping, rounded, dark green leaflets which are pubescent and biserrate, with small, dark, reddish-brown glands on the teeth and by the large bipinnate sepals, reflexed and concealing the hips (though the sepals fall early). Prickles too, are often a characteristic feature of *R. obtusifolia*; although strong they are not coarse, the inner radius of the prickle forming an almost complete semicircle whilst a section at the base of the prickle tends to be circular rather than elliptical. The flowers are occasionally pale pink but this probably indicates some introgression of *R. canina*.

R. obtusifolia 14

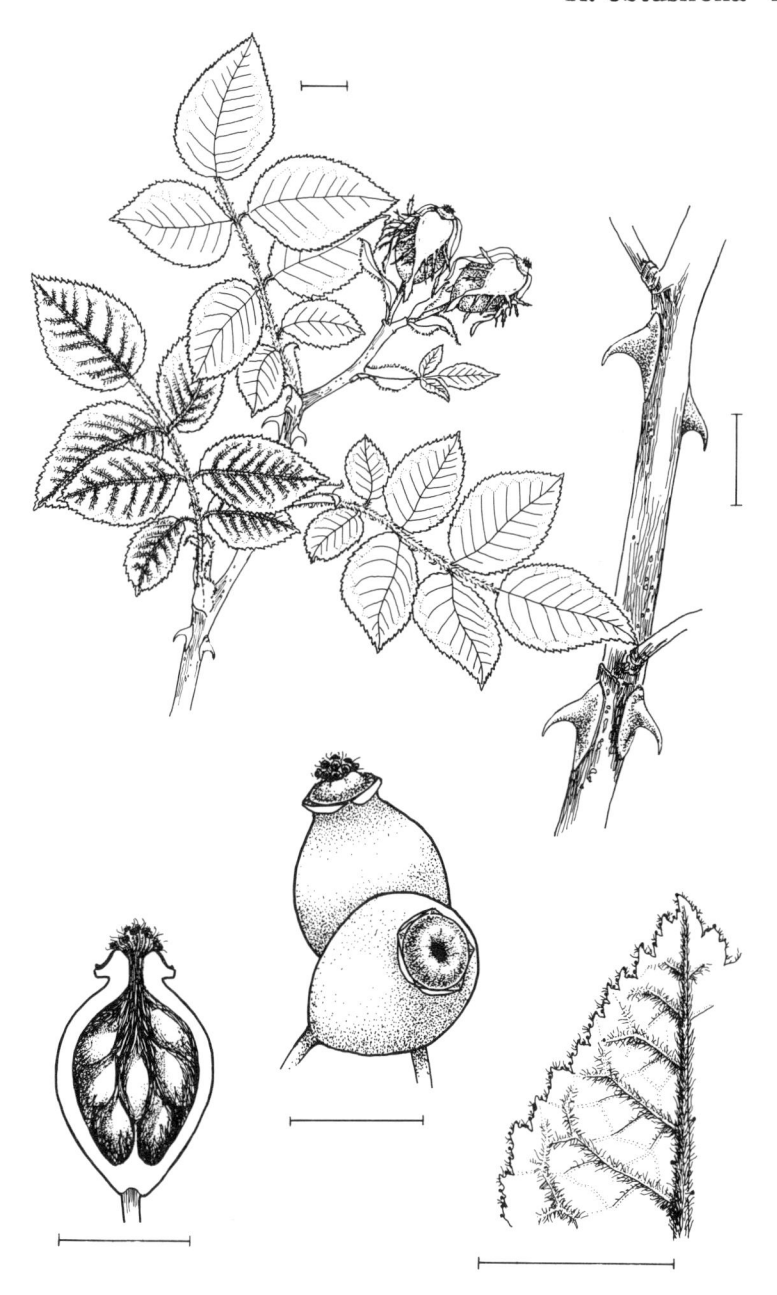

14 × 4. Rosa obtusifolia Desv. × R. arvensis Huds.
= R. × rouyana Duffort ex Rouy

General habit of *R. obtusifolia*, with young stems wine-red on sun-lit side. Prickles are numerous, stout, strongly curved, and occurring in clusters so that the whole bush is fiercely prickly. Leaflets are narrower, larger and more acute than are those of *R. obtusifolia*, sparsely pubescent beneath, irregularly biserrate, with some poorly developed glands on the denticles. Main teeth show signs of the curvature of those of *R. arvensis*. Hips are 0.5-1.5cm, globose, the smaller somewhat misshapen and perhaps partially sterile; pedicels may be 2.5cm, glandular-hispid; the sepals are bipinnate and strongly reflexed; the stigmas are hispid, the styles not exserted.

A rare hybrid of which we have seen very few specimens.

V.-c. 37,55.
Non-directional records of the hybrid from v.-c. 31,42.

14 × 11. Rosa obtusifolia Desv. × R. stylosa Desv.

One living specimen and two herbarium sheets only have been seen. They differ from *R. obtusifolia* in having slightly coarser prickles, and glandular-hispid pedicels. Some of the fruits have low, though undoubtedly conical discs with clumped styles.

A rare hybrid, possibly overlooked as it is difficult to determine.

V.-c. 1,29.
Non-directional records of the hybrid from v.-c. 35.

14 × 12. Rosa obtusifolia Desv. × R. canina L.
= R. × dumetorum Thuill. Map 19

Where these two species occur together there is often a bewildering series of hybrids showing all gradations between the two species. It is therefore virtually impossible to give satisfactory descriptions which would delimit the reciprocal hybrids. The influence of *R. obtusifolia* is usually easy to observe, but the assumption of *R. canina* influence is often a matter of probability, based on knowledge of the roses of a particular region. When the *R. canina* influence is only slight, e.g. if the only departure from *R. obtusifolia* consists of slightly larger flowers with pale pink petals, the specimen can be recorded as *R. obtusifolia*, with a small degree of permissible introgression. *R. canina* with suspected traces of *R. obtusifolia* influence is best ignored for recording purposes, as it is too near *R. canina* Group *Pubescentes*. Where there is strong evidence of the presence of both species the specimen should be labelled *R.* × *dumetorum*.

V.-c. 1,17,22,55,58.
Non-directional records of the hybrid from v.-c. 4,6,12,13,15,18,21,23, 25-27,29,31-38,40-43,49,50,57,60-64,H18,19,21,26,40.

14 × 13. Rosa obtusifolia Desv. × R. caesia Sm.

Low-growing shrub with the neat and compact appearance of *R. obtusifolia*; young stems are strongly wine-red on the sunlit side. Leaflets resemble those of *R. obtusifolia* in general appearance, but are subglabrous with only slight glandular biserration. Pedicels are short but not concealed by bracts. Sepals are spreading-erect, bipinnate. Stigmas are in a broad villous head.

A rare hybrid as neither parent is abundant in regions where their ranges overlap. It is not always clear which is the female parent.

V.-c. 55.
Non-directional records of the hybrid from v.-c. 57,58.

14 × 15. Rosa obtusifolia Desv. × R. tomentosa Sm.

Determined for H. Handley by Melville but not seen by us.

V.-c. 55.
Non-directional records of the hybrid from v.-c. H14.

14 × 18. Rosa obtusifolia Desv. × R. rubiginosa L. = R. × tomentelliformis W.-Dod

Nearer to *R. obtusifolia* in habit and armature. Acicles are found on pedicels and on some flowering branches. Typical Sweet-briar-type glands are found on the leaves beneath and occasionally on the sepals.

A specimen we have examined from NMW, distributed by the Botanical Exchange Club as representative of this hybrid, is an indeterminate *Rosa rubiginosa* hybrid with little resemblance to *Rosa obtusifolia*. However, three further specimens from BM and CGE seem to us to be of the hybrid described above.

V.-c. 32,58.

15. Rosa tomentosa Sm.   Map 26

Harsh Downy-rose

Climbing shrub up to 3m with arching stems; young stems green; prickles arcuate, slender but strong. Leaflets 5-7, (1.5-)2-4 × 1-2cm, ovate-lanceolate, acute, irregularly biserrate with small secondary teeth, pale green or greyish green, tomentose especially beneath, usually with faintly resin-scented or odourless glands on lower surface (which may be hidden by the tomentum); petiole and rachis tomentose and glandular; stipules 10-15 × 4-7mm, densely gland-fringed, with spreading auricles. Flowers 1-6, 4cm in diameter, pink or occasionally white. Hips 1-1.5cm × 1-1.2cm, globose or ovoid, red when ripe, glandular-hispid, at least in the lower part; pedicels long, 2-3.5cm, densely glandular-hispid; sepals 1.5-2cm, narrow, with a slightly expanded tip and a few lateral lobes, densely glandular all over, spreading or spreading-erect after anthesis but falling before the hips have fully reddened; styles glabrous or slightly hispid; stigmas in a small head; orifice small, 1/5 diameter of disc or less. $2n = 35$(unbalanced). Flowering June-July.

With its climbing habit this is mainly a hedgerow species, but it also occurs at the edge of woodland and in scrub and especially in scrub on chalk in the South. Southern in distribution, becoming rare in the North, and then usually occurring as isolated bushes.

R. tomentosa can be distinguished from the other Downy-roses by its climbing habit, long pedicels, small stylar orifice and the early-falling sepals. However, where it occurs together with *R. sherardii*, especially in the North-West, there are often puzzling intermediates which are best left to a specialist for determination. There is a slight difference in gland type between the two species in that those of *R. tomentosa* are distinctly less odorous, less translucent and often less conspicuous. In the South, and in parts of Ireland, especially on chalk, it frequently occurs together with *R. micrantha*, which is similar in general appearance but quite different in gland type. In some areas hybrids with *R. canina* are frequent and such hybrids are determined only by paying close attention to the small differences noted under *R. × scabriuscula*.

R. tomentosa 15

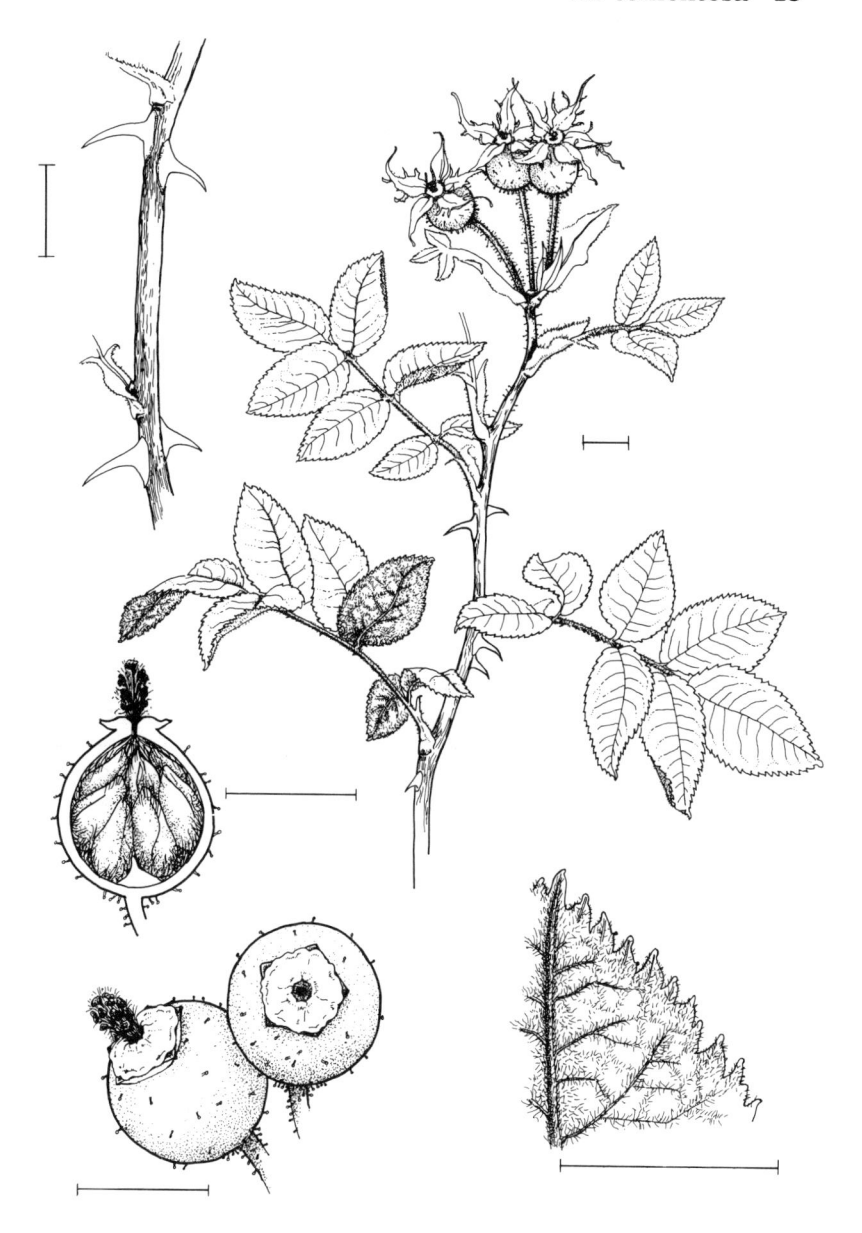

15 × 5. Rosa tomentosa Sm. × R. pimpinellifolia L.
= **R. × coronata** Crépin ex Reuter

General habit of *R. tomentosa* and only separated from it by the very mixed armature and occasional presence of smaller, more glabrous leaflets.

In areas where the ranges of *R. tomentosa* and *R. sherardii* overlap their hybrids with *R. pimpinellifolia* may be difficult or impossible to separate. Careful comparison should be made with the local populations of the putative parent species.

V.-c. 10.
Non-directional records of the hybrid from v.-c. 49,50,H39.

15 × 12. Rosa tomentosa Sm. × R. canina L.
= **R. × scabriuscula** Sm.   Map 20

Habit and general appearance of *R. tomentosa*. Leaflets are sparsely pubescent, sometimes on veins only beneath, weakly biserrate with a few glands on the teeth, and glands rare or absent on lower surface. Sepals are reflexed and fall early. Styles and stigmas are often glabrous.

This is virtually *R. tomentosa* with certain features suppressed or reduced in quantity. As *R. canina* has no outstanding features, its role as male parent seems to be that of suppression, various key-features of the female parent being apparently much reduced rather than replaced.

V.-c. 3,17,23,31,32,36,37,41,44,49,53,55,57,58,64,H38.
Non-directional records of the hybrid from v.-c. 5-8,11,12,15,16,19,20, 22,26,28,29,35,38,40,42,43,48,63,65,H2,4,5,10,11,14,16,17,19-21,33, 40.

15 × 13. Rosa tomentosa Sm. × R. caesia Sm.
= R. × rogersii W.-Dod

Often has 7 leaflets, which are elliptic to ovate, rather strongly biserrate and glabrous, or nearly so, except for glands. The rachis is glandular-hispid and has numerous prickles. The hips are ellipsoidal and often infertile and the pedicels are slender and glandular-setose.

The male parent is *R. caesia* subsp. *glauca* according to Melville, but his description might just as well apply if subsp. *caesia* were involved. Again the description does not rule out *R. sherardii* as female parent, and this would be much more likely in northern and Scottish vice-counties. We have seen only rather poor herbarium specimens.

V.-c. 37.

15 × 14. Rosa tomentosa Sm. × R. obtusifolia Desv.

Habit of *R. tomentosa*, tall and arching with green stems; prickles are mostly strongly curved as in *R. obtusifolia* but with some considerably less curved and more slender on the upper parts of the stem. Leaflets are intermediate in shape, larger than in *R. obtusifolia* but more rounded in outline than in *R. tomentosa*, irregularly multiserrate with *R. obtusifolia*-type glands on the smaller teeth, almost glabrous above, rather thinly pubescent with scattered glands on the midrib and veins beneath. Petiole and rachis are tomentose, with stipitate glands. Hips are abortive, small and misshapen; pedicels are glandular hispid; the sepals are strongly reflexed, pinnate, some lobes having very small secondary lobes.

V.-c. 55.
Non-directional records of the hybrid from v.-c. H14.

15 × 16. Rosa tomentosa Sm. × R. sherardii Davies = R. × suberectiformis W.-Dod

The two species are so closely related that in the few areas where they grow together in any quantity hybrids appear intermediate and it is almost impossible to determine the female parent.

We have seen plants 2-3m, with strong, arching stems, pedicels (1-)2-2.5cm, sepals falling early, stylar orifice 1/4 diameter of disc, but otherwise matching the description of *R. sherardii* × *R. tomentosa*. These would appear to have *R. tomentosa* as female parent.

V.-c. 46,70,88.
Non-directional records of the hybrid from v.-c. H19.

15 × 17. Rosa tomentosa Sm. × R. mollis Sm.

General habit of *R. tomentosa*, climbing to 2m or more, strong and arching. However the stem armature includes several long, straight, patent prickles, some indeed rising above the horizontal. Some leaflets are rather wide for *R. tomentosa* and softly tomentose beneath. Hips are very variable, some to 2.5 × 1.5cm, with short pedicels 1-1.5cm in length, others are smaller with longer pedicels; sepals are erect to spreading as the hips ripen; stigmas are tomentose and the orifice 1/3-1/2 the diameter of the disc.

V.-c. 59.
Non-directional records of the hybrid from v.-c. 39.

15 × 18. Rosa tomentosa Sm. × R. rubiginosa L.
= R. × avrayensis Rouy

Has the tall, strong, climbing habit of *R. tomentosa* with a mixture of arcuate, declining and strongly arching prickles on the main stem. A few acicles are often present on the flowering branches. Leaflets are mostly ovate or elliptic, rather long, sparsely pubescent on both surfaces, with a few Sweet-briar-type glands beneath. Pedicels are long, glandular-hispid, some with acicles and viscid glands. Hips are narrowly ovoid, some being aborted.

A rare hybrid of which we have seen only two specimens.

Non-directional records of the hybrid from v.-c. 35,44.

15 × 19. Rosa tomentosa Sm. × R. micrantha Borrer ex Sm.

Tall, climbing and arching, with strong, arcuate, broad-based prickles. Leaflets are ovate, moderately large, glandular-biserrate, pubescent beneath, with *R. tomentosa*-type glands on the midrib and Sweet-briar-type glands on the lower surfaces. Hips are narrowly ovoid or ellipsoid; sepals are spreading or slightly reflexed, soon falling.

The tall, arching habit and lack of strongly curved prickles and acicles differentiate this from the hybrid with *R. rubiginosa*.

V.-c. 17,46.
Non-directional records of the hybrid from v.-c. 12.

15 × 20. Rosa tomentosa Sm. × R. agrestis Desv.

Very similar to the hybrid with *R. micrantha*, but with appreciably smaller leaves which are narrowed at the base (but hardly cuneate), shorter pedicels which are only sparsely glandular-hispid, and almost glabrous stigmas.

The single specimen seen was growing in an area where *R. agrestis* is reasonably plentiful and *R. micrantha* is absent.

V.-c. H10.

16. Rosa sherardii Davies   Map 27

Sherard's Downy-rose

Erect shrub 1-2m, with slender, somewhat glaucous or occasionally wine-red pigmented stems, often zig-zag in lower parts and flexuous at the extremities; prickles normally slender, arcuate-acuminate or declining with weak bases (stronger, more curved prickles are prevalent in some regions). Leaflets 5-7, 2.5-3.5 × 1.5-2cm, medium-sized, ovate or broadly elliptical, somewhat rugose, bluish-green and sparsely tomentose above, light-green or ashy and softly tomentose beneath (though sometimes only on the veins), with many or no resinous subfoliar glands, glandular-multiserrate; petiole and rachis pubescent and glandular; stipules 15-20 × 8-10mm, wide, gland-fringed. Flowers 1-4, 2.5-3.5cm in diameter, deep rose-pink (often white in Scotland). Hips 1.5-2.5 ×(0.8-)1-1.5cm, glandular-setose, globose, obovoid, narrowly ovoid to ellipsoid (with regional variation), ripening fairly early, red when ripe; pedicels 1-1.5cm, glandular-hispid; sepals 1.5-2cm, pinnate, very glandular with reddish, rather translucent, resin-scented glands, spreading-erect after anthesis, slightly constricted at the attachment with the hip, but persisting until the hips are almost ripe; styles villous; stigmas in a large, domed, very hispid head covering 2/3 of the disc; disk more or less flat with a rounded lip to the orifice which is 1/3 diameter of disc (slightly less in southern forms). $2n = 28,35,42$(unbalanced). Flowering June-July.

Woodland margins, open scrub moorland and mountain tracks. Frequent in Scotland, northern England and parts of Wales; decreasing southwards and largely replaced by *R. tomentosa* in southern England.

R. sherardii is easily recognized as one of the Downy-roses by its hairy leaflets and glandular pedicels. In habit, size, length of pedicels, position and disarticulation of the sepals and size of stylar aperture *R. sherardii* is intermediate between *R. tomentosa* and *R. mollis*. There appears to be more regional variation in *R. sherardii* than in most British wild roses. Fortunately several key characters, such as spreading-erect sepals, large stylar orifice and dome-shaped head of hairy stigmas, are characteristic of this species in all regions.The glands of the Downy-roses differ in size, odour and appearance from those of the Sweet-briars. When fresh they vary in appearance from brown-red and slightly viscid to translucent. They are subsessile and have a resinous odour when crushed. At 0.1-0.15mm in diameter they are only half the size of Sweet-briar glands but both dry out to a greyish crust in the herbarium.

R. sherardii 16

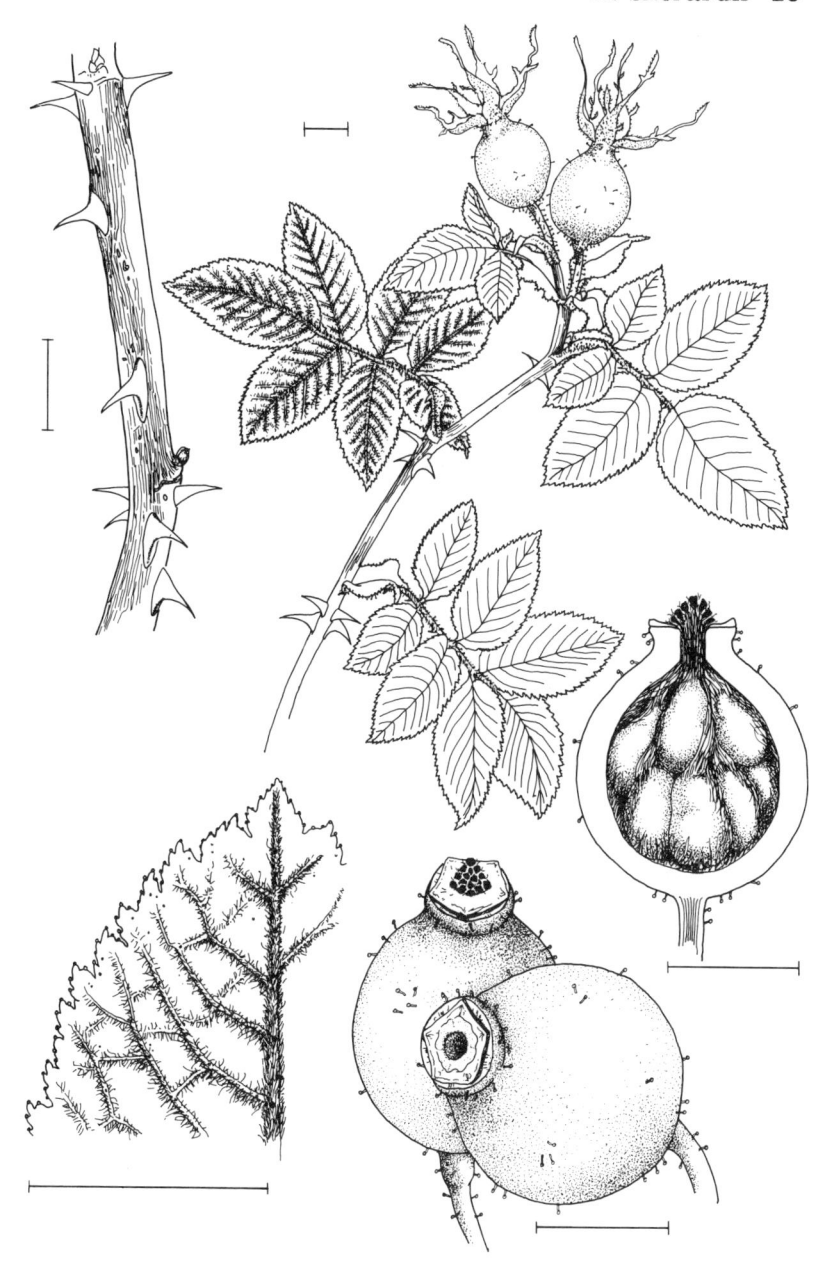

121

16 × 4. Rosa sherardii Davies × R. arvensis Huds.

General habit of *R. sherardii* but distinguished by the presence of long flexuous leading shoots, tinged with wine-red. Leaflets are glandular-biserrate with small closely-set teeth, bluish green, slightly rugose and subglabrous above, tomentose beneath. Petiole and rachis are strongly glandular. Hips are ovoid, smaller than those of *R. sherardii*, glandular-hispid all over. Sepals are spreading-erect, persistent, lobed and glandular, but somewhat shorter than those of *R. sherardii*. Pedicels measure 2-2.5cm, longer than is usual in *R. sherardii* and are strongly glandular-hispid; stigmas are in a small head, almost glabrous.

A rare hybrid with only three records.

V.-c. 42,48,55.

16 × 5. Rosa sherardii Davies × R. pimpinellifolia L. = R. ×involuta Sm. Map 7

A tall bush, slender and arching, with a very mixed armature of strong acicles and slender, arcuate prickles. Leaflets are 0.9-1.7 × 1.5-3cm, variable in size and shape on the same bush, some orbicular with less tomentum than *R. sherardii*, fewer glands and intermediate serration, others are more glandular and ovate. Hips are occasionally ill-formed and sterile. Sepals are subpinnate, some falling early.

There have been many opinions as to the identity of *R.* × *perthensis* Rouy, first discovered at Auchterarder, mid Perth, by W. Barclay in 1892. The resulting correspondence and subsequent literature on the subject are fully discussed by J.R. Matthews (1934,1976). Matthews (1934) raised the plant from 'seed' and reported that all the offspring were alike. He was convinced that it was not a hybrid but "a unique form of '*R. omissa*'" (a very glandular form of *R. sherardii*). Melville (1975) considered the plant to be a triple hybrid, *R. pimpinellifolia* × *R. sherardii* × *R. rubiginosa*, listing characters which seemed to him to support this conclusion. We have had the opportunity to examine several herbarium specimens and a fresh specimen, kindly sent from Auchterarder by Dr R.F. Thomas, but could find no conclusive signs of *R. rubiginosa* in this material. The specimens indeed showed extensive acicular development on both hips and pedicels but this is not necessarily an indicator of *R. rubiginosa*. On the other hand the presence of viscid, sweet-scented glands, a more diagnostic feature of *R. rubiginosa*, could neither be affirmed nor denied as the herbarium specimens were old and dry and the fresh specimen had been collected too late in the season. Further investigation is required. For the time being we can only conclude that *R.* × *perthensis* is a rather abnormal form of *R. sherardii* × *R. pimpinellifolia* with numerous glands on the leaflets and excessive acicular development of hips and pedicels, with *R. sherardii* as female parent.

V.-c. 1,17,38,49,64,79,88,H105,108.
Non-directional records of the hybrid from v.-c. 37,42,44,48,50,52,57, 62,66-70,82,84,89,90,94-96,100,108,110,H8,12,14,20,21,26,35,38,39.

16 × 12. Rosa sherardii Davies × R. canina L. = R. × rothschildii Druce   Map 21

Has low, arching stems, usually only 1.5m in height; prickles are slender, inclined or arcuate, interspersed with others more broad-based, curved and rather blunt. Leaflets are large, with few or no glands, some being almost glabrous and uniserrate. Hips are variable on the same bush, larger than is usual in *R. sherardii*, a few are very small and appear sterile; pedicels are 1.5cm (occasionally 2cm) in length.

This hybrid is virtually *R. sherardii* with reduced characters so that it is difficult to determine if the mixed armature is not apparent.

V.-c. 6,26,35,37-39,41,42,44,47,49,55,57,64,67,69,70,83,100,H20,21.
Non-directional records of the hybrid from v.-c. 36,43,45,46,48,50-52,58,62,66,82,85,94,96,H6,8,12,14,18,19,22,23,29-32,35,36,38,39.

16 × 13. Rosa sherardii Davies × R. caesia Sm.   Map 24

General form of *R. sherardii*, but with stronger stems, larger hips, and some reduction in glandulosity and tomentum of the leaflets. Pedicels are often short, with fewer glands than is usual in *R. sherardii*.

This hybrid can only be identified with any certainty by paying close attention to the local rose populations. Strongly wine-red-pigmented stems and leaves, hips with spreading-erect sepals, a large stylar-aperture and a large villous head of stigmas distinguish the hybrid from *R. sherardii* × *R. canina*. The short pedicels (1-1.5cm) help to separate it from *R. tomentosa* × *R. caesia*.

V.-c. 66,72,105,108.
Non-directional records of the hybrid from v.-c.42,52,59,69,70, 82,83,88,92,95,96,98-100,103,105-107,H22,29,32,33,35,38,39,40.

16 × 15. Rosa sherardii Davies × R. tomentosa Sm.
= R. × suberectiformis W.-Dod

The two species are so closely related that in the few areas where they grow together in any quantity hybrids appear intermediate and it is almost impossible to determine the female parent. The following description is taken from field-notes on a score or so of bushes which appear to have *R. sherardii* as female parent.

Stems 1.5-2m, climbing, somewhat arching, slender; prickles are arcuate-acuminate, some slender and some robust. Leaflets are 4-5 × 2-3cm, sparsely tomentose above, tomentose and sometimes glandular beneath; teeth are irregularly biserrate, with three small, faintly resinous glands per tooth; stipules are 15 × 3-4mm. Hips are 2.5 × 1.5cm, ovoid; pedicels are 1-2(-2.5)cm, glandular-hispid; sepals are speading-erect, falling as the fruit ripens; stigmas are hispid in a large head; the orifice is 1/3 diameter of disc.

V.-c. 59,69.
Non-directional records of the hybrid from v.-c. 46,H19.

16 × 17. Rosa sherardii Davies × R. mollis Sm.
= R. × shoolbredii W.-Dod   Map 28

General habit of *R. sherardii* with perhaps more erect stems and a tendency to suckering. Slender and arcuate prickles are interspersed with others which are patent and completely straight. Large, softly-tomentose leaflets are often found along with others more rugose and hispid on the same bush. There is often an appreciable neck of tissue between the medium-sized hips and the semi-persistent, subpinnate sepals.

Although these species grow in mixed populations in Scotland and northern England, hybrids between them are rare and are not easily determined. Most of the criteria which separate the two species relate to size and degree rather than to form and are best estimated in the field, close attention being paid to the local populations of the parents.

V.-c. 46,69.
Non-directional records of the hybrid from v.-c. 35,41,42,44,65,66,82, 84,88,95,106,111,H31,32.

16 × 18. Rosa sherardii Davies × R. rubiginosa L.
= R. × suberecta (Woods) Ley

Habit closely resembling *R. sherardii*, but stout, very curved prickles are present among the generally slender, arcuate armature of the main stems. Leaflets are small to medium, ovate, with rounded bases, rugose, tinged with wine-red, glandular-biserrate as well as glandular beneath. Glands on leaflets, stipules and sepals are mixed, both resinous and Sweet-briar-type being present. Hips are medium-sized, subglobose to obovoid, glabrous or with a few acicles. Pedicels are more or less glandular-hispid, occasionally aciculate.

V.-c. 55,69,79,89,105,106,108.
Non-directional records of the hybrid from v.-c. 51,59,82,84,90,95,96, H4,17,29,31,33,39.

16 × 19. Rosa sherardii Davies × R. micrantha Borrer ex Sm.

A more or less erect shrub with the general habit of *R. sherardii*, or with some arching stems, but distinguished by the presence of viscid, Sweet-briar-type glands at least on the sepals and pedicels and, sometimes, on the leaves and stipules. Armature mixed, with rather coarse, strongly curved prickles and more slender armature on the same stem; acicles absent from all parts of the plant. Hips variable, some with an appreciable neck, rather small stylar orifice and hispid stigmas, others with more characters of *R. sherardii* including erect, late-falling sepals.

V.-c. 42,46,H3.

16 × 20. Rosa sherardii Davies × R. agrestis Savi

General habit of *R. sherardii*. Prickles are few, mixed, mostly slender and slightly curved, together with a few that are coarser and broad-based. Leaflets are broadly elliptic to cuneate, rugose, sparsely hispid above, sometimes more hispid beneath. Glands referable to both parent species are present on leaflets and rachis. Flowers are pink. Hips measure 2 × 1.3cm and are obovoid. Pedicels measure 0.8-1.2cm and are short and glandular, with reddish glands which spread to the hips. Sepals are short, with narrow pinnae, spreading-erect, but falling as the hips ripen. Stigmas are in a large head, hispid. Orifice is 1/4 diameter of disc.

V.-c. H10,29.

17. Rosa mollis Sm. Map 29

Soft Downy-rose

Erect shrub 0.8-1.5(-2)m, suckering and, when conditions are favourable, forming dense thickets; main branches straight; well lit young stems suffused with wine-red; shaded stems glaucous; prickles slender, quite straight, patent (often rising above the horizontal). Leaflets 3-5(-7), 2-4 × 1.5-2.5cm, large, oblong, ovate or elliptical, subobtuse, irregularly glandular-serrate, softly tomentose on both surfaces, ashy-grey beneath, the tomentum hiding numerous, reddish, subsessile glands, the glands when crushed releasing a resinous odour; petiole and rachis glandular-hispid; stipules 15-20 × 7-12mm, wide, gland-fringed. Flowers 1-3, 3-4.5cm in diameter, usually deep pink though pure white forms occur. Hips 1.5-3 × 1.5-2cm, large, broadly ellipsoid, red when ripe (Scottish forms may be globose and purplish when ripe), often glandular-hispid; pedicels 0.5-1(-1.5)cm, short, glandular-hispid; sepals 2-2.5cm, more or less simple, expanded into small or large, leaf-like tips, erect and persistent until the hip decays, attached to the hip by a thick ring of tissue and not constricted at the attachment; styles villous; stigmas in a large, hispid, hemispherical head, almost concealing the slightly concave disc; orifice very wide, 1/2 diameter of disc or more (rarely less). $2n = 35, 42$(unbalanced). Flowering June-July.

Woods, hedges, waste ground, ascending high into the hills. Frequent in Scotland and northern England, becoming rarer southwards. Local in Ireland and mainly in the North. One of the earliest of our native roses to flower and fruit.

R. mollis is distinguished by its strong, upright, suckering stems with straight internodes and completely straight, patent prickles. From *R. pimpinellifolia*, the only other British wild rose with a similar habit, it is separated by its large hairy leaflets and red hips. From the other Downy-roses it is separated by its large, softly-tomentose leaflets, large hips with very short pedicels, erect, more or less simple sepals persisting until the hips decay, and a very large stylar orifice. Very old plants may reach 2m, with stems 10cm in thickness, branching only at the apex and leaning at an angle from the vertical but with no sign of arching.

R. mollis 17

127

17 × 5. Rosa mollis Sm. × R. pimpinellifolia L.
= R. × sabinii Woods
Map 8

In habit like *R. mollis*. Stems have numerous patent prickles and acicles of varying size. There are 5-7 leaflets, large and softly pubescent, glands being sparse beneath or absent. Hips are smaller than in typical *R. mollis*, but otherwise resemble them; they are sometimes sterile. Pedicels are mostly short, but occasionally reach 2.5cm, and are sparsely glandular-hispid.

V.-c. 62,64,65,66,69,80,89,92,94,108,111.
Non-directional records of the hybrid from v.-c. 16,37,57,67,88,90,91, 95,96,100,106,107,109,H9,39.

17 × 12. Rosa mollis Sm. × R. canina L.
= R. × molletorum H.-Harr.

General habit of *R. mollis*, but with rather broad-based, curved prickles on the main stems amongst the main armature of straight, patent prickles. Leaflets are glabrous above, with very sparse tomentum beneath. Hips are smaller and of varying shapes. Sepals reach an erect position on the ripening fruit but fall early.

Scarcely to be separated from *R. mollis* × *R. caesia*.

V.-c. 39,57.
Non-directional records of the hybrid from v.-c. 60,66,67,68,104.

17 × 13. Rosa mollis Sm. × R. caesia Sm.
= R. × glaucoides W.-Dod

General habit strong and upright as in *R. mollis* though main stems sometimes branching at a lower node; often found in large suckering colonies. Leaflets may be less glandular than in *R. mollis* and only slightly hispid. Sepals often have leaf-like lobes and are persistent on the ripe hips. Difficult to separate from the hybrid with *R. canina*, the balance of probability in Scotland being for *R. caesia* as male parent.

V.-c. 57,97.
Non-directional records of the hybrid from v.-c. 66-68,70,95,96,111.

17 × 16. Rosa mollis Sm. × R. sherardii Davies
= R. × shoolbredii W.-Dod   Map 28

Differs from *R. mollis* in only a few salient features. Arcuate prickles occur amongst a preponderantly straight, patent armature. Leaflets often vary greatly in size, some at least having the proportions of *R. sherardii*. Hips are a good guide as they are usually large, sometimes having pedicels up to 2cm in length, and appearing top-heavy so that a few become almost pendulous. Sepals may be slightly pinnate and constricted at their junction with the hip.

Although these species grow in mixed populations in Scotland and northern England, hybrids between them are rare and are not easily determined. Most of the criteria which separate the two species relate to size and degree rather than to form and are best estimated in the field, close attention being paid to the local populations of the parents.

V.-c. 65,67,69,96,105,106.
Non-directional records of the hybrid from v.-c. 35,41,42,44,46,66,82, 84,88,95,111,H31,32.

17 × 18. Rosa mollis Sm. × R. rubiginosa L.
= R. × molliformis W.-Dod

General habit erect and suckering with stems branching only at the apex. Strongly curved prickles of various sizes may usually be found among the more slender, patent armature typical of *R. mollis*. Leaflets vary in character, some are large, ovate and densely pubescent, with reddish, sessile glands beneath, others are smaller and almost orbicular, with scattered Sweet-briar-type glands beneath. Brownish, viscid glands may be found on petiole, rachis and stipules. Hips are very variable, some measuring 1 × 1.2cm, others 2.2 × 1.7cm. Sepals are erect on the ripe fruit and may be gland-fringed.

V.-c. 91.
Non-directional records of the hybrid from v.-c. 80,82,88,90,95,96,109.

18. Rosa rubiginosa L. Map 30

Sweet-briar, Eglantine

Erect shrub 1-2m; stems strong; prickles strong, very curved, acuminate, unequal and often mixed with scattered stout acicles, particularly on the flowering branches below the inflorescence. Leaflets 5-7, 1-2(-2.5) × 1-2cm, suborbicular to ovate-elliptical, strongly glandular-biserrate, pubescent on the veins beneath and slightly so above, with many viscid, brownish or translucent, sweet-smelling glands beneath; petiole and rachis glandular; stipules 12 × 5-8mm, gland-fringed. Flowers 1-3, 3-4cm in diameter, bright pink. Hips 1.2-2.5 × 1-1.5cm, ovoid, obovoid or subglobose, scarlet when ripe, smooth, or glandular-setose at base; pedicels c.1cm, glandular-hispid, the glands often mixed with small acicles; sepals 2cm, slightly pinnate, glandular, spreading-erect after flowering, usually persisting until the hips redden; stigmas hispid, in a wedge-shaped head, sunk in the wide orifice; disc shallowly concave; orifice c. 1/3 diameter of disc. $2n = 35$(unbalanced). Flowering June-July.

Occurs mainly on calcareous soils; with its upright and self-supporting habit it is particularly characteristic of open scrub on chalk or limestone, but is also found in hedgerows. Where soils are suitable it occurs throughout Britain and Ireland, but is commonest in the South. As it is now frequently introduced, the status of some populations is doubtful.

The combination of upright habit, neat glandular-biserrate leaflets, with numerous, conspicuous, subfoliar glands and glandular-hispid pedicels will usually distinguish this species in the field. The presence of very curved, acuminate, unequal prickles and possibly some acicles are useful confirmatory characters for *R. rubiginosa* and its hybrids, but acicles may sometimes be absent. The larger prickles, especially on the upper parts of the stem, are often rather long and shaped like an eagle's talon. Those lower down on the stem may be very strongly curved, shorter and with the point turned vertically down like an eagle's beak.

To many people *R. rubiginosa* advertises its presence on a warm, still day by the ripe-apple scent of its leaves. However others with a keen sense of smell can detect no odour at all so this character should not be treated as diagnostic until one has confirmed one's ability to detect it. In the herbarium the very viscid, translucent glands of all the Sweet-briars and their hybrids may dry out to a whitish crust and are then difficult to distinguish from glands of the Downy-roses, which undergo a similar change. Nevertheless, if subfoliar in position, the two types of glands can often be distinguished, those of the Sweet-briars being evidently stalked, and those of the Downy-roses more or less sessile and often hidden among the leaflet tomentum.

R. rubiginosa 18

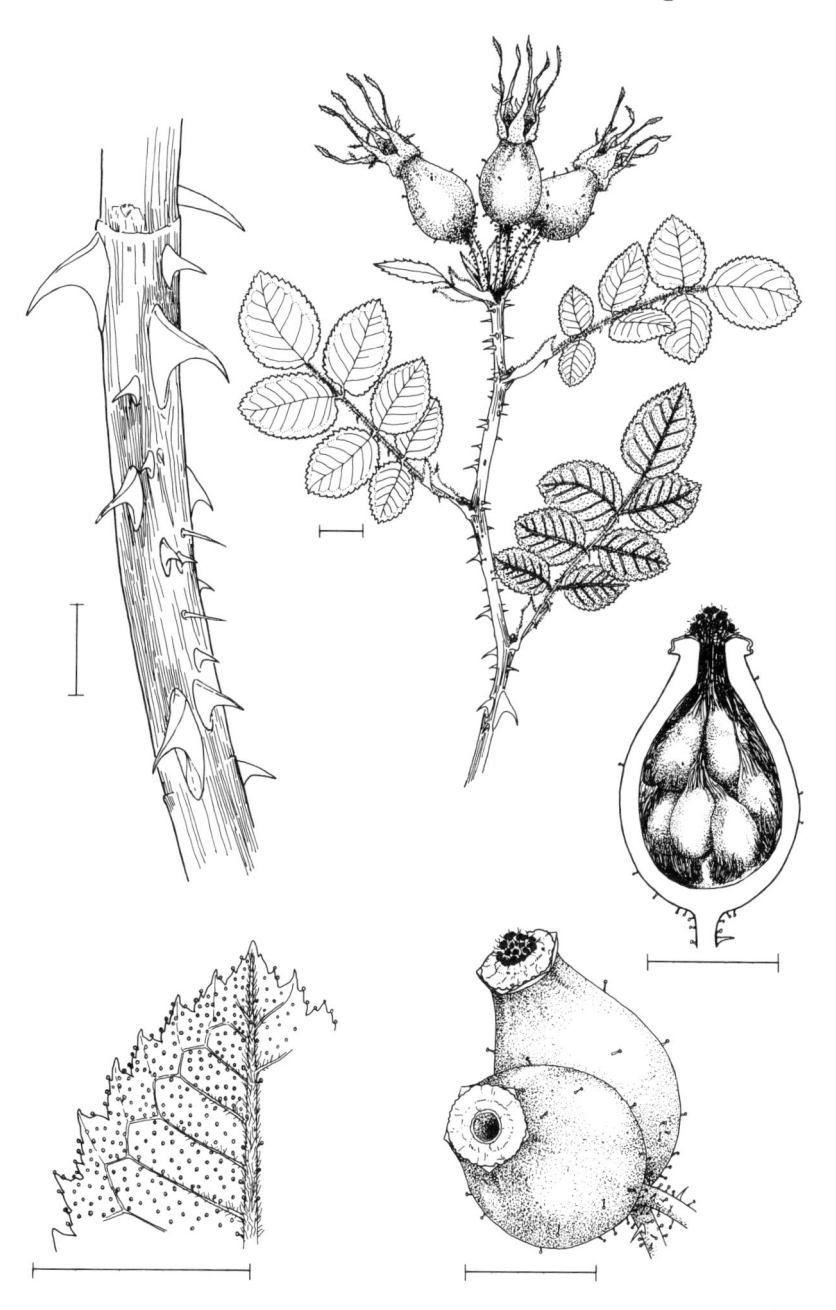

18 × 4. Rosa rubiginosa L. × R. arvensis Huds.
= R. × consanguinea Gren.

A somewhat weaker bush than typical *R. rubiginosa* with young shoots tending towards the trailing habit of *R. arvensis*. Leaflets are smaller than those of *R. rubiginosa*, the glandular biserration of the teeth being irregular or reduced. Hips are often aborted. No directional material has been seen and the description is taken from Melville (1955) who recorded it for v.-c. 44.

Non-directional records of the hybrid from v.-c. 16.

18 × 5. Rosa rubiginosa L. × R. pimpinellifolia L.
= R. × cantiana (W.-Dod) W.-Dod   Map 9

General habit of *R. rubiginosa*; armature is mixed, the strong, arched prickles of *R. rubiginosa* clearly standing out amongst the slender prickles and acicles of *R. pimpinellifolia*. Leaflets have scattered Sweet-briar-type glands beneath, but approach *R. pimpinellifolia* in size and shape. Hips are mostly solitary, small and rather urceolate; sepals fall early.

We have seen two herbarium sheets only of this hybrid.

Non-directional records of the hybrid from v.-c. 15,57,59,64,66,67,79, 80,82,88-91,95,106,H16,26,38.

18 × 11. Rosa rubiginosa L. × R. stylosa Desv.

Habit open and more straggling than is usual in *R. rubiginosa*, with rather widely spaced prickles and some reduction in numbers of pricklets. Some leaflets ovate and rather large, but with Sweet-briar-type subfoliar glands. Hips small, ovate, slightly urceolate, but with fairly thick conical discs or aborted; pedicels short, but some up to 2cm; styles few, more or less glabrous, ending in an ill-formed head of stigmas.

A rare hybrid of which we have seen few specimens.

V.-c. 33.

18 × 12. Rosa rubiginosa L. × R. canina L.
= R. × nitidula Besser

Habit strong and arching with occasional thick-based, rather blunt prickles among the characteristic strongly curved, acuminate prickles of *R. rubiginosa*. Leaflets ovate or ovate-lanceolate, occasionally subglabrous, with sharply biserrate teeth, viscid glands sparse or wanting. Hips globose, ellipsoid or narrowly ellipsoid, some partially sterile, ripening late; pedicels sometimes smooth; sepals falling well before the hips ripen; stylar orifice often only 1/4-1/3 diameter of disc.

This hybrid is in essence *R. rubiginosa* with certain features such as glandulosity somewhat reduced; whether it can therefore be separated from *R. rubiginosa* × *R. caesia* is uncertain as the latter hybrid has not yet been recorded. Our notes are taken from Gustaffson (1944) and Melville (1975).

V.-c. 29.
Non-directional records of the hybrid from v.-c. 16,17,28,29,31,38,41, 42,55,57,64,66,H39.

18 × 15. Rosa rubiginosa L. × R. tomentosa Sm.
= R. × avrayensis Rouy

Close to *R. rubiginosa* in habit and armature though some slender, arcuate prickles can also be found on the main stems. Leaflets are much larger than in typical *R. rubiginosa* and are glabrous above, pubescent and with Sweet-briar-type glands beneath. Hips are small, subglobose, with rather long pedicels which are beset with small reddish glands as well as with a few acicles. The sepals fall early.

A rare hybrid of which we have seen only two specimens.

Non-directional records of the hybrid from v.-c. 35,44.

18 × 16. Rosa rubiginosa L. × R. sherardii Davies
= R. × suberecta (Woods) Ley

Very close to *R. rubiginosa* and only separated from it by the mixed armature of the main stem where slender, arcuate prickles are usually found amongst more strongly arched prickles typical of *R. rubiginosa*. Some leaves have only few Sweet-briar-type glands. Pedicels may or may not be glandular-aciculate.

A difficult hybrid to recognise, the reciprocal being more commonly recorded.

V.-c. 55.
Non-directional records of the hybrid from v.-c. 51,59,82,84,90,95,96, 106,108,H4,17,29,31,33,39.

18 × 17. Rosa rubiginosa L. × R. mollis Sm.
= R. × molliformis W.-Dod

Habit of *R. rubiginosa*, but stems have a mixed armature. Long, completely straight, patent prickles typical of *R. mollis* stand out amongst otherwise strongly curved prickles. Leaflets are suborbicular, medium in size, glandular-multiserrate, densely pubescent on both surfaces, with numerous glands beneath. Hips are subglobose and sparsely glandular-setose; pedicels are short, glandular and aciculate; the sepals are suberect and glandular; styles and stigmas villous; the orifice is 1/2 diameter of the disc.

A rare hybrid noted only from Scotland.

V.-c. 110.

Non-directional records of the hybrid from v.-c. 80,82,88,90,95,96,109.

19. **Rosa micrantha** Borrer ex Sm. Map 31

Small-flowered Sweet-briar

Tall shrub 1.5-3m, climbing and arching; stems green, with long internodes; prickles very curved, sparse, with long bases, more or less equal; acicles absent. Leaflets 5-7, 1.5-3(-3.5) × 0.8-2.5cm, ovate, obovate or elliptical, rounded at base, usually glabrous above, with viscid, brownish, or translucent sweet-smelling glands and short pubescence beneath, glandular-multiserrate; petiole and rachis glandular; stipules 15 × 4mm, fringed with viscid glands. Flowers 1-4, 2-3.5cm in diameter, pink. Hips 0.9-1.7 × 0.7-1cm, small to medium, urceolate or with a short neck, red when ripe; pedicels 1-1.7(-2)cm, nearly always glandular, but with no acicles; sepals 2cm, pinnate, reflexed, falling before the hips ripen; styles glabrous, slightly exserted; stigmas in a subglobose head; disc convex; orifice 1/6-1/5 diameter of disc. $2n = 35,42$(unbalanced). Flowering June-July.

Woods, scrub and hedges; mainly, but not always restricted to calcareous soils. Widespread in southern England, rare in the North (probably now extinct in Durham) and very rare in Scotland.

R. micrantha is known immediately as one of the Sweet-briars by the sweet-scented, viscid, brownish or translucent glands which are present on leaflets, petioles, stipules, pedicels and sepals. It is separated from *R. rubiginosa* by its taller climbing habit, more equal prickles, absence of acicles from all parts of the bush, longer leaflets, smaller flowers and hips, reflexed sepals and more glabrous styles. In more open habitats where it may resemble *R. rubiginosa* in general habit it can be seen to have weaker stems with longer internodes.

R. micrantha 19

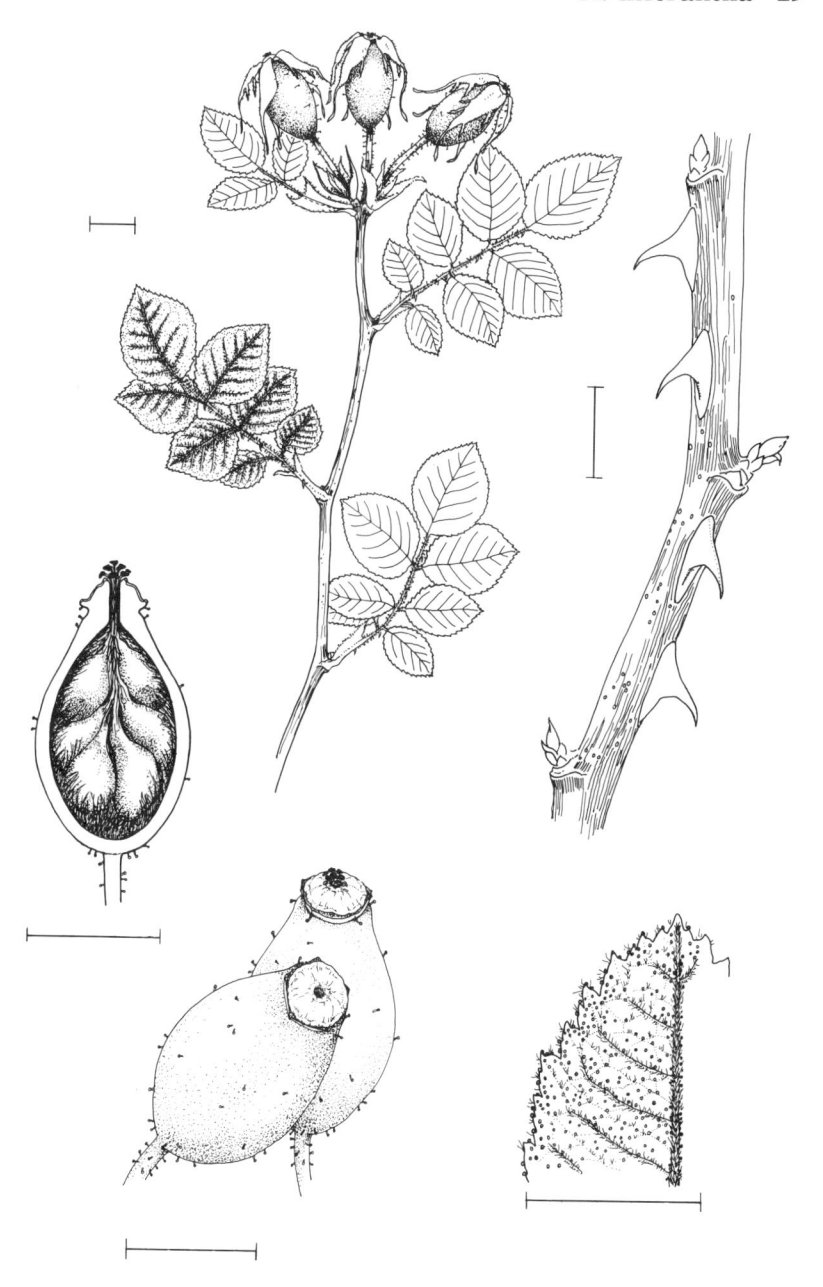

19 × 4. Rosa micrantha Borrer ex Sm. × R. arvensis Huds. = R. × inelegans W.-Dod

General habit of *R. micrantha*, with perhaps longer and weaker leading stems, strongly tinged with wine-red. Prickles resemble those of *R. micrantha*. Other characters showing the influence of *R. arvensis* are extremely variable, even on the same bush, and are not easy to detect. Leaflets may be rather small, pale grey-green glaucous beneath, with teeth strongly resembling those of *R. arvensis*, except that some may be glandular. Scattered Sweet-briar-type glands are also found on most leaflets beneath. Hips seem to vary considerably in size and shape, some resembling *R. arvensis* with flat disc, others tending towards the urceolate shape of *R. micrantha* with shallowly convex disc; some hips may be aborted; pedicels are long and glandular-hispid; styles are somewhat exserted, with varying degrees of fusion.

V.-c. 23,32.
Non-directional records of the hybrid from v.-c. 48,H5.

[19 × 11. Rosa micrantha Borrer ex Sm. × R. stylosa Desv.]

Doubtfully recorded; we have as yet seen no examples.

19 × 12. Rosa micrantha Borrer ex Sm. × R. canina L. = R. × toddiae W.-Dod

Habit of *R. micrantha*, with tall, slender, strongly climbing stems and rather sparse, strongly curved prickles. Leaflets are of intermediate shape, glabrous above, sparsely glandular and pubescent beneath. Similar Sweet-briar-type glands may also be found on stipules and sepals and sometimes also on the pedicels, but more sparsely so than in typical *R. micrantha*. Hips are larger than is usual in *R. micrantha*, some being obovoid, others aborted. Sepals are reflexed, falling early.

V.-c. 3,17,23,25,37.
Non-directional records of the hybrid from v.-c. 5,11,23,35,36,38,41, 42,48,49,H3.

19 × 14. Rosa micrantha Borrer ex Sm. × R. obtusifolia Desv.

Stems with widely spaced leaves and long-based, very curved prickles. Leaflets very similar to those of *R. obtusifolia* in dentition, shape, and indumentum but with scattered viscid glands beneath. Hips 1.4 × 0.8-1cm and 1.5 × 1.3cm, mixed ellipsoid-urceolate and subglobose; pedicels 1-1.3cm, short, glandular; sepals reflexed, some bipinnate, falling before the hips ripen; head of stigmas rather open and slightly hairy.

A rare hybrid of which, until recently, only one rather unsatisfactory herbarium specimen, lacking prickles, was available for examination. We have since seen a specimen from Somerset which, in our opinion, represents this hybrid.

V.-c. 6

19 × 18. Rosa micrantha Borrer ex Sm. × R. rubiginosa L. = R. × bigeneris Duffort ex Rouy

Habit tall and arching like *R. micrantha*, but with acicles on the flowering branches; prickles are strongly curved and unequal. Leaflets are broadly ovate or suborbicular. Hips are generally small and urceolate with glabrous or hispid stigmas; some are more globose with rising sepals and moderately large orifice.

According to Melville the reciprocal hybrid has not been recorded: it would in any case be very difficult to determine.

V.-c. 17.
Non-directional records of the hybrid from v.-c. 15,16,28.

19 × 20. Rosa micrantha Borrer ex Sm. × R. agrestis Savi = R. × bishopii W.-Dod

Armature stout, curved and rather widely spaced on the stems. Leaflets are small, with scattered Sweet-briar-type glands beneath; some, at least, have a cuneate base. Hips are larger than in typical *R. micrantha*; pedicels are c. 1cm, sparsely glandular-hispid.

A rare combination as *R. agrestis* is itself rare. Melville (1975) lists only the reciprocal hybrid. However, the six rather unsatisfactory specimens we have critically examined appeared to us to have *R. micrantha* as female parent.

V.-c. 9,11,13,17.

20. Rosa agrestis Savi Map 32

Small-leaved Sweet-Briar

A medium sized, erect, free-standing shrub 1-1.5(-2)m; young stems somewhat flexuous, green; prickles curved, stout-based. Leaflets 5-7, 2-2.5 × 1-1.3cm, narrowly elliptical, a large proportion cuneate at base and acute at apex, glandular-multiserrate with scattered, viscid or translucent, scarcely sweet-smelling glands beneath; petiole and rachis glandular; stipules 10-15 × 5-6mm, gland-fringed. Flowers 1-4, 2-4cm in diameter, white or pale pink. Hips 1-2 × 1-1.5cm, ovoid to globose or slightly urceolate, red when ripe, smooth; pedicels 1-1.5cm, smooth; sepals 2cm, pinnate, reflexed, falling early; styles glabrous or slightly hairy; disc slightly convex; orifice 1/6-1/5 diameter of disc. $2n = 35$(unbalanced). Flowering June-July.

Scrub on calcareous soils, now very rare in southern England and Wales, absent from Scotland, and local and rare in Ireland. This species is easily separated from *R. rubiginosa*, as it lacks most of the salient features of that species; from *R. micrantha* it is differentiated by its smooth pedicels and high proportion of narrow leaflets which are cuneate at the base.

R. agrestis 20

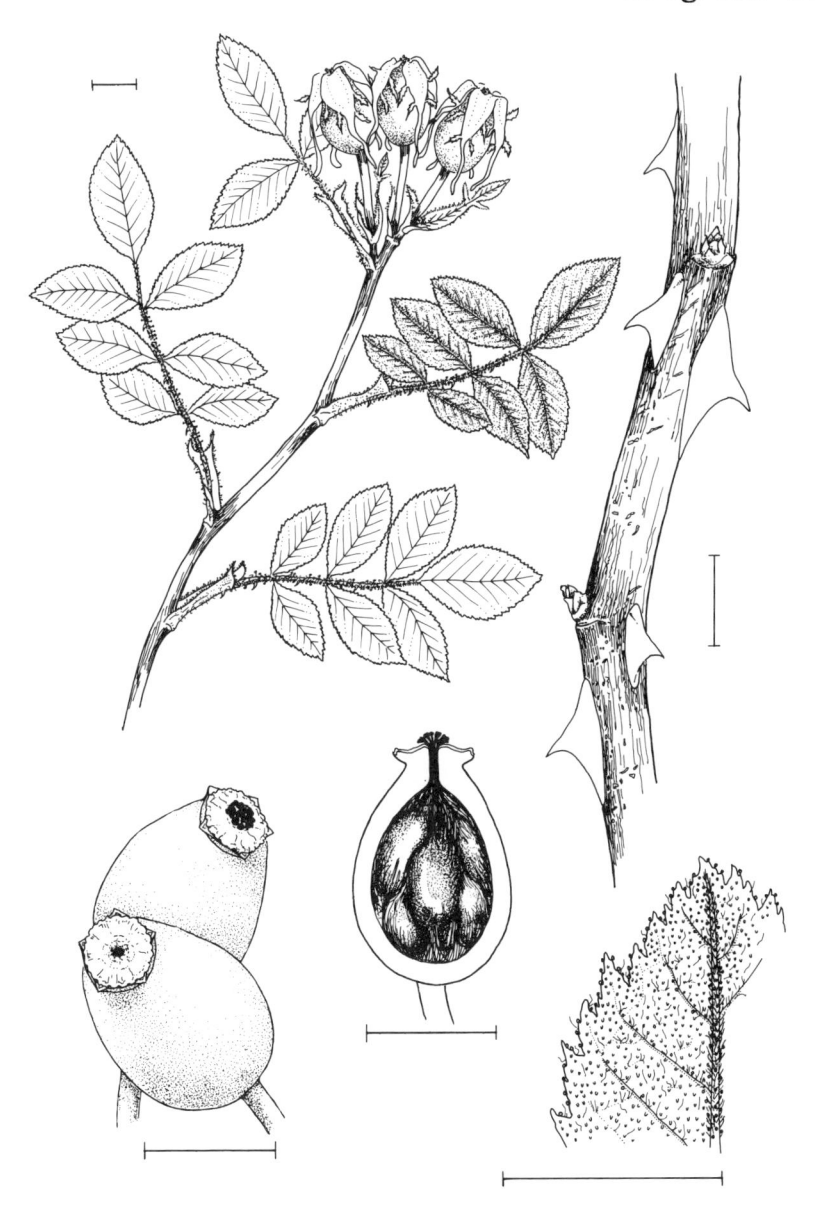

20 × 11. Rosa agrestis Savi × R. stylosa Desv.

General habit of *R. agrestis*. Differs in the leaflets which are often very mixed, some are glandular, rather large and elliptical, others are cuneate at the base yet with few glands. The hips normally resemble those of *R. agrestis* but some have a reduced conical disc and a slightly exserted, rather plump column of styles can usually be found.

A rare hybrid of which only two examples have been seen.

V.-c. 6.

20 × 12. Rosa agrestis Savi × R. canina L.
= R. × belnensis Ozan

General habit of *R. canina* with large, broad-based prickles. Leaflets are intermediate, some cuneate-based and moderately glandular, others with more rounded bases and fewer glands; the glands closely resemble those of *R. agrestis* populations nearby. Hips are of differing sizes, some are small and slightly urceolate, others are much larger and subglobose; pedicels measure 1.5-2cm and are rather long; sepals have rather slender lobes, are sparsely glandular and fall early.

A very rare hybrid of which only two specimens have been seen by us.

V.c. 27.

[20 × 16. Rosa agrestis Savi × R. sherardii Davies]

Melville recorded this hybrid from Ballyvaughan, Ireland, but the only specimens we have seen appear to us to have *R. sherardii* as female parent.

[20 × 19. Rosa agrestis Savi × R. micrantha Borrer ex Sm. = R. × bishopii W.-Dod]

Noted by Melville (1975) for three Vice-counties. The material we have critically examined appeared to us to be the reciprocal hybrid.

12. Distribution maps

Source of the records

Two of the native British and Irish roses, *R. arvensis* and *R. pimpinellifolia* are relatively easy to identify. Their distribution was mapped in the *Atlas of the British Flora* (Perring & Walters 1962). The maps of these species in this Handbook are based on records collected for the Atlas by the B.S.B.I. Distribution Maps Scheme (1954-9), augmented by records subsequently sent to the Biological Records Centre, those made in the B.S.B.I. Monitoring Scheme (1987-8) and records from the sources used for the more critical species, outlined below. The map of the genus and that of the alien species *R. rugosa* (which was not mapped in the Atlas) were prepared from the same sources of data.

Mapping the remaining native taxa has been less straightforward, as these species and their hybrids are much less easy to identify. No attempt has previously been made to map even the species at the 10km square scale. The Atlas mapped only three broadly defined aggregates, and no attempt was made to cover *Rosa* in its *Critical Supplement* (Perring & Sell 1968). It was not until 1975 that Melville's account of Rose hybrids laid the foundation for further research on a sound taxonomic basis, and it is not surprising therefore that many county Floras in recent years are woefully inadequate in their treatment of the genus.

The maps of the native critical species and the more frequent hybrids presented below have therefore been prepared from data compiled specially for the purpose. Records of the rarer aliens and hybrids have been collected in a similar way, but summarized as vice-county records in the text. Only records which we have good reason to believe are based on accurate determinations have been included.

The main sources of records for the maps of the critical taxa are:

1. Specimens in the herbarium of A. H. Wolley-Dod (BM). This contains some 3000 specimens which have recently been redetermined by the authors (cf. Primavesi 1992).

2. Specimens in ABRN, CGE, DBN, E (including herb. H.H. Johnston), LIV, LTR, NMW and the private herbarium of Professor G.A. Swan. Specimens in ABRN, CGE, DBN, LIV, LTR and herb. G.A.S. have been redetermined by A.L.P. Specimens in NMW determined by experts according to Wolley-Dod's taxonomic system were accepted for those taxa which correspond to specimens recognized here and the remainder determined by A.L.P. Records from E are also based on expertly determined specimens of taxa which correspond to the species recognized here; the remainder have not been examined.

3. Specimens collected for the uncompleted British Rose Survey (1952-3), organized by R. Melville and P. Sylvester-Bradley. These specimens are now housed in LTR and were determined by A.L.P. (cf. Gornall & Primavesi 1989).
4. Specimens collected by D.A. Doogue during a recent intensive survey of Irish roses and determined or authorized as acceptable by A.L.P.
5. Records collected during the B.S.B.I. Distribution Maps and Monitoring Schemes, and those submitted to the Biological Records Centre since the publication of the *Atlas of the British Flora*. All these have been vetted by A.L.P. and only localized, expertly determined records have usually been accepted.
6. Records acquired by A.L.P. in his capacity as B.S.B.I. referee for *Rosa*, and his personal field records.
7. Records from the recently published Floras of Co. Durham (Graham 1988), Leicestershire (Primavesi & Evans 1988), Moray, Nairn and East Inverness (McCallum Webster 1978) and Rutland (Messenger 1971).
8. Records collected during recent fieldwork in Cornwall and Devon (L. J. Margetts), Fife (G. H. Ballantyne), Cumbria (G. Halliday et al.) and Worcestershire (R. Maskew et al.).

These records, if not already on computer file, have been computerized by the Biological Records Centre, Institute of Terrestrial Ecology, and are held in the ORACLE database management system. The maps have been prepared from this database by BRC, to whom grateful acknowledgement is made.

The maps of the critical species and the hybrids must be regarded as provisional. It has not been possible to include all records which are potentially available; problems of distance and time have, for example, prevented the authors from extracting records from some of the major herbaria including BM (excluding herb. Wolley-Dod) and K. Even if all available records had been obtained, coverage would still have been patchy and in some areas quite inadequate. Comparison of the map of all roses (Map 1) with the maps of the segregates reveals how inadequate is the coverage. It is probably least satisfactory for the Groups of *R. canina* which were not distinguished by Clapham et al. (1952, 1962, 1987) and therefore only became known to most recorders on the publication of Stace (1991). Nevertheless, a start on mapping the *Rosa* taxa has now been made. It is hoped that the distribution maps, together with the taxonomic accounts in the Handbook, will encourage more people to take an interest in the genus and stimulate further recording.

Records are mapped in the 10 × 10km squares of the Ordnance Survey National Grid in Great Britain and in the Ordnance Survey/Suirbheiracht Ordonais National Grid in Ireland. Records from the Channel Isles are mapped in the 10 × 10km squares of the Universal Transverse Mercator Grid. The symbols used are:

- ■ records made in or after 1950
- ▲ records made before 1950, or undated
- × introduction (used only for records of species that are native in other parts of Britain and Ireland).

Supplementary bibliography

CLAPHAM, A.R., TUTIN, T.G. & WARBURG, E.F. 1952. *Flora of the British Isles.* Cambridge.

CLAPHAM, A.R., TUTIN, T.G. & WARBURG, E.F. 1962. *Flora of the British Isles.* Ed. 2. Cambridge.

CLAPHAM, A.R., TUTIN, T.G. & MOORE, D.M. 1987. *Flora of the British Isles.* Ed. 3. Cambridge.

GRAHAM, G.G. 1988. *The flora and vegetation of County Durham.* Durham.

McCALLUM WEBSTER, M. 1978. *Flora of Moray, Nairn & East Inverness.* Aberdeen.

MESSENGER, K.G. 1971. *Flora of Rutland.* Leicester.

PERRING, F.H. & SELL, P.D. (eds) 1968. *Critical supplement to the Atlas of the British Flora.* London.

PERRING, F.H. & WALTERS, S.M. (eds) 1962. *Atlas of the British Flora.* London & Edinburgh.

Map 1. **Rosa** : all records

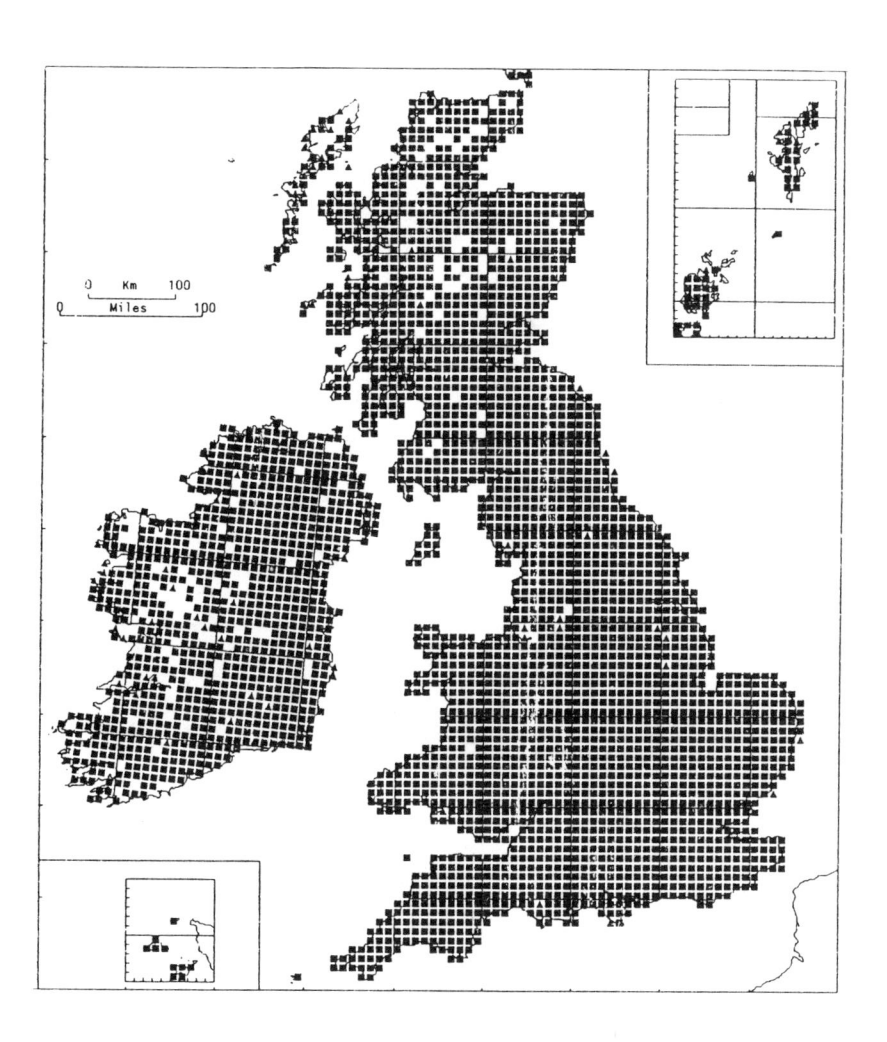

Map 2. **R. arvensis 4**

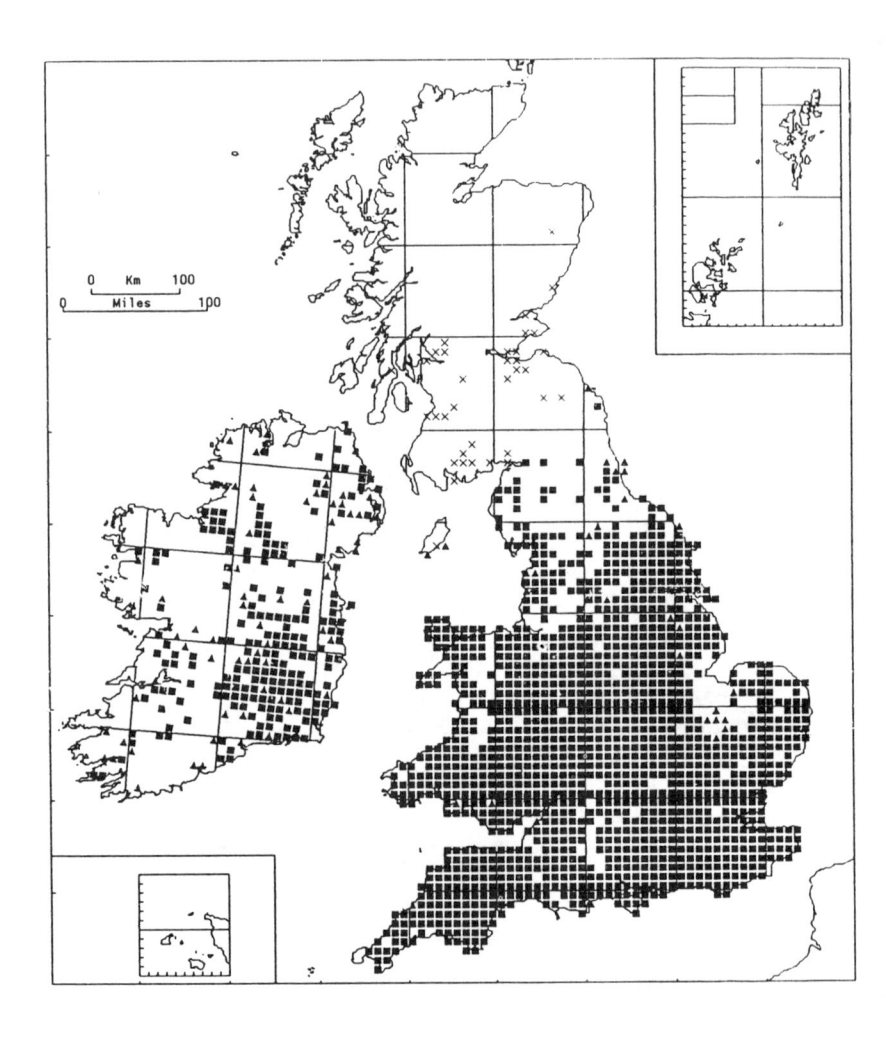

Map 3. **R. × pseudorusticana**
  (**R. arvensis** × **R. stylosa**) 4 × 11

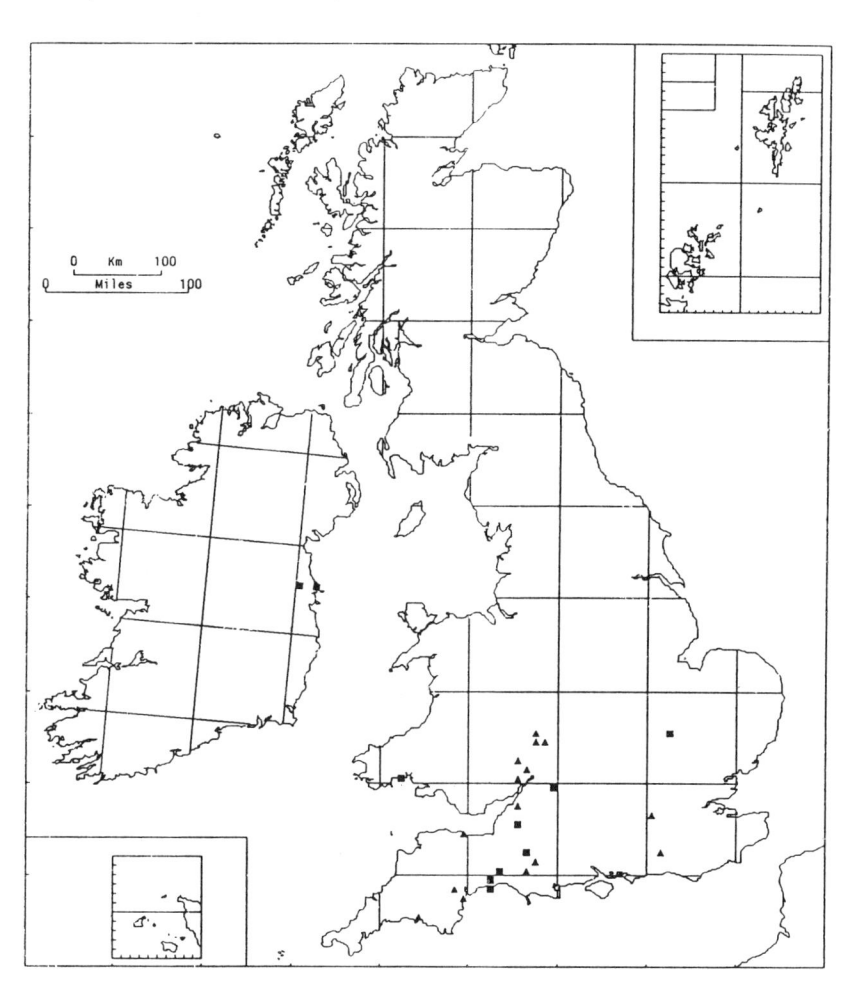

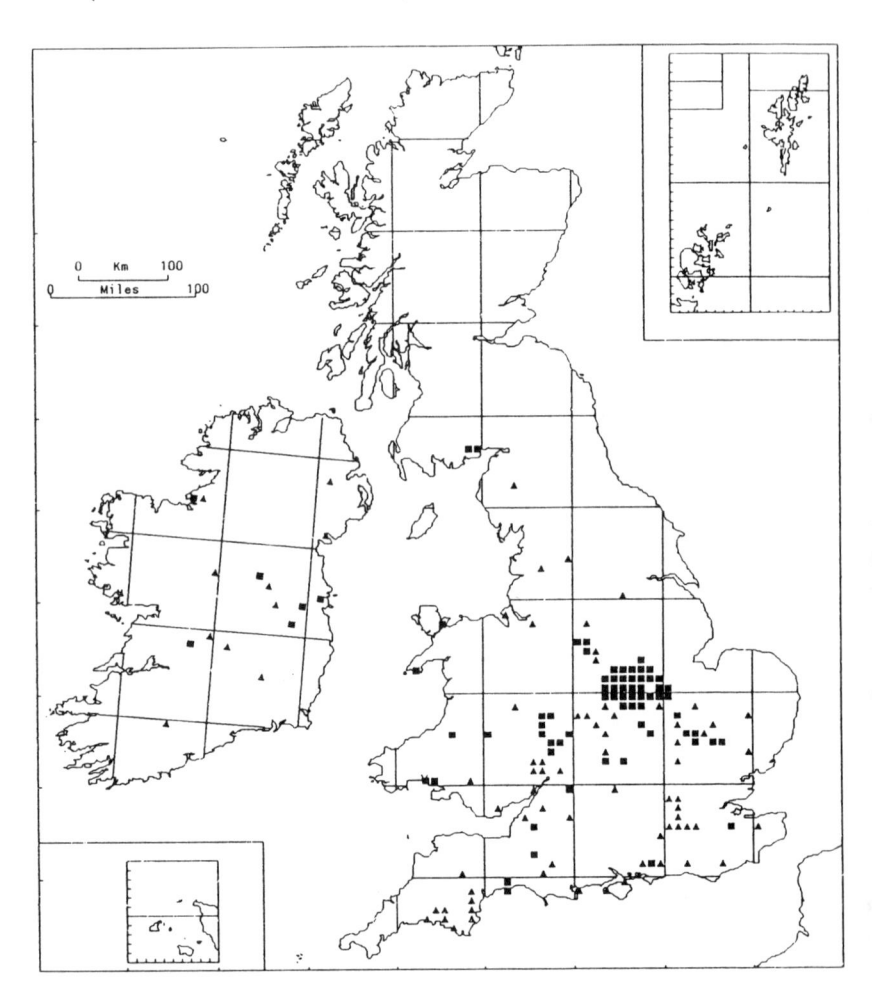

Map 4. R. x **verticillacantha**
(**R. arvensis** x **R. canina**) 4 x 12

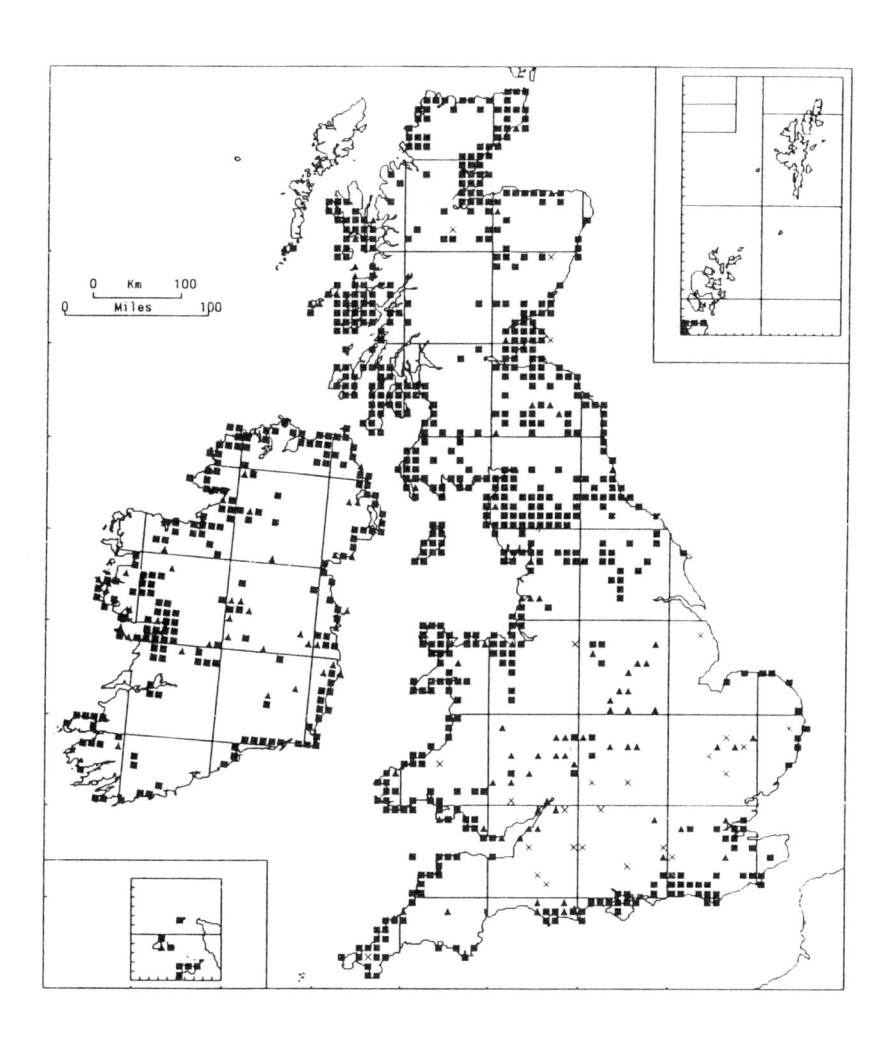

Map 5. **R. pimpinellifolia 5**

Map 6. **R. x hibernica**
(**R. canina** x **R. pimpinellifolia**) 12 x 5

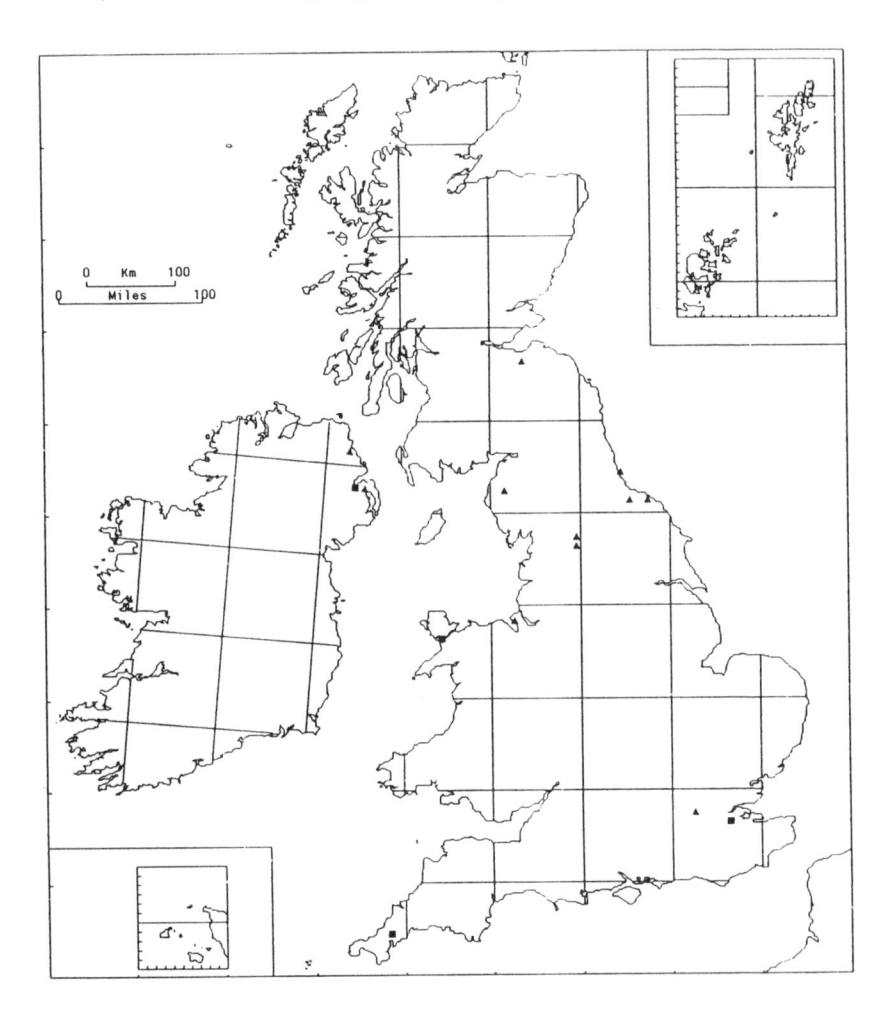

Map 7. **R. × involuta**
 (**R. pimpinellifolia** × **R. sherardii**) 5 × 16

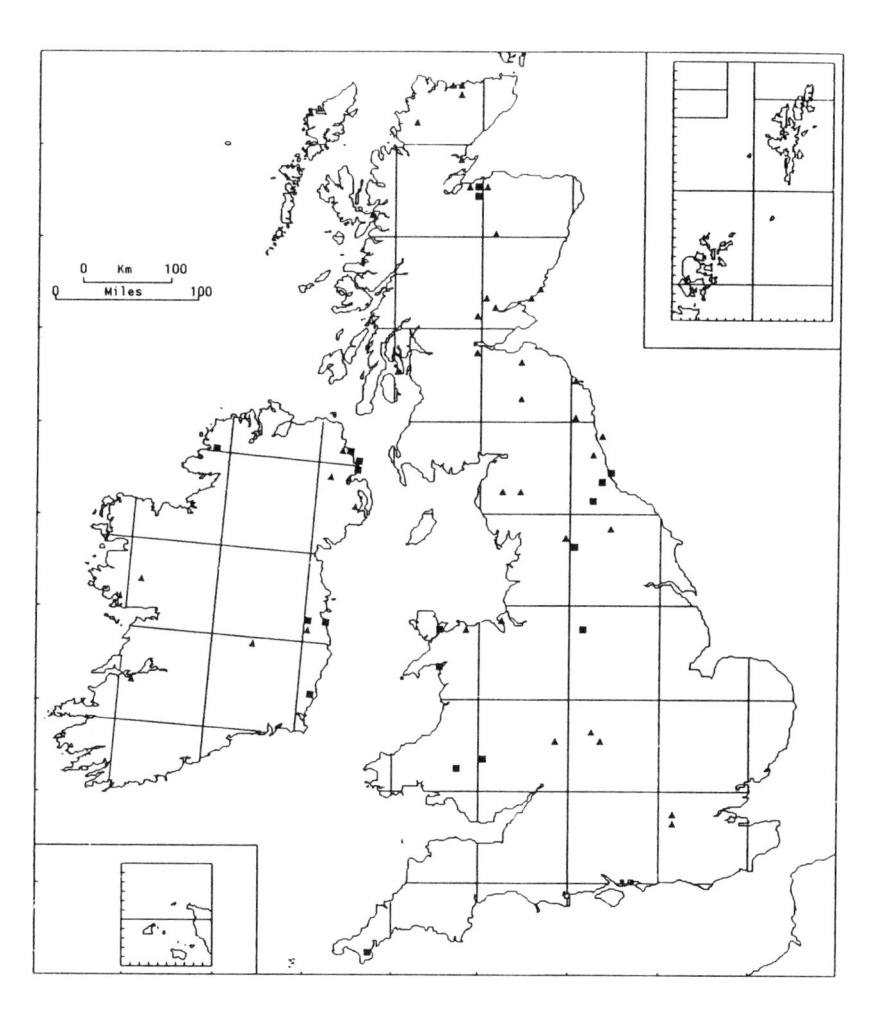

Map 8. **R. x sabinii**
(R. mollis x R. pimpinellifolia) 17 x 5

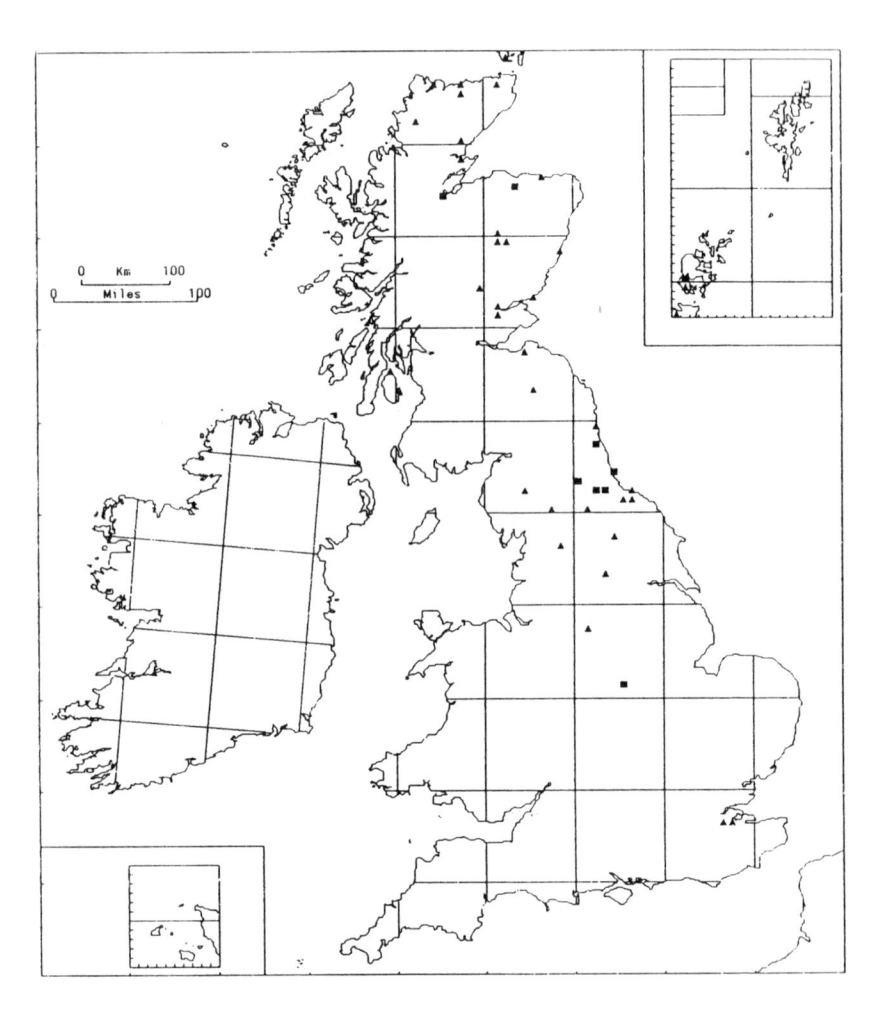

Map 9. **R.** x **cantiana**
 (**R. pimpinellifolia** x **R. rubiginosa**) 5 x 18

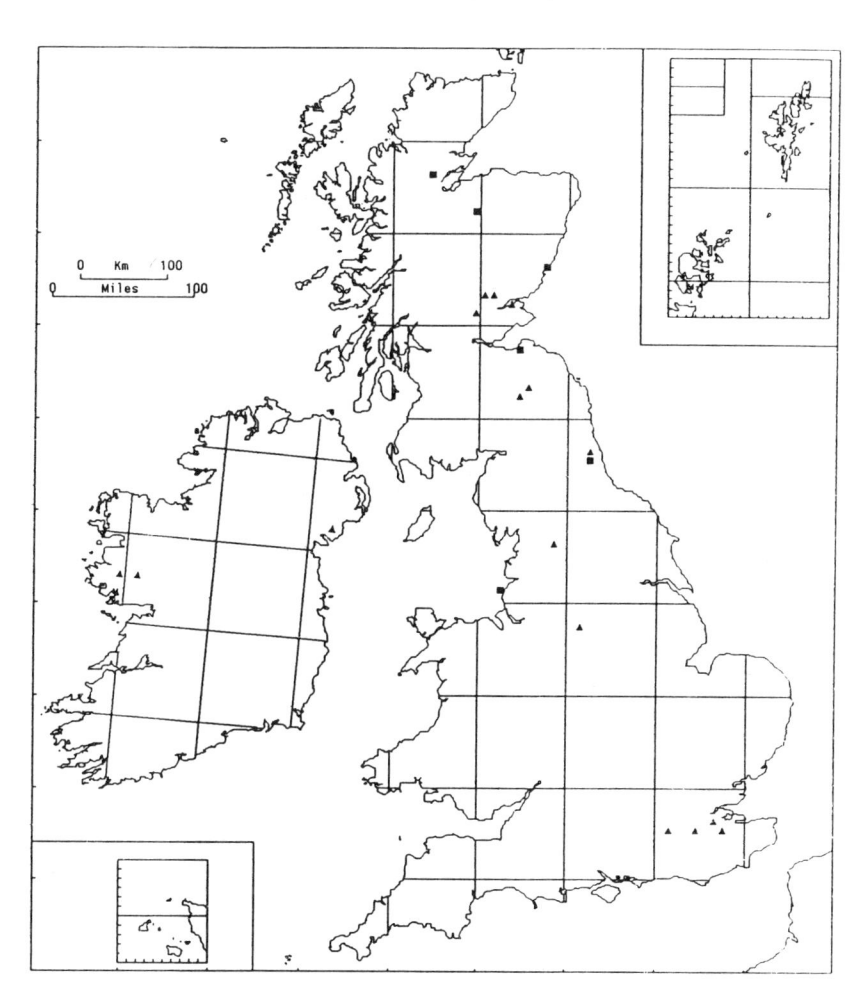

Map 10. **R. rugosa 6**

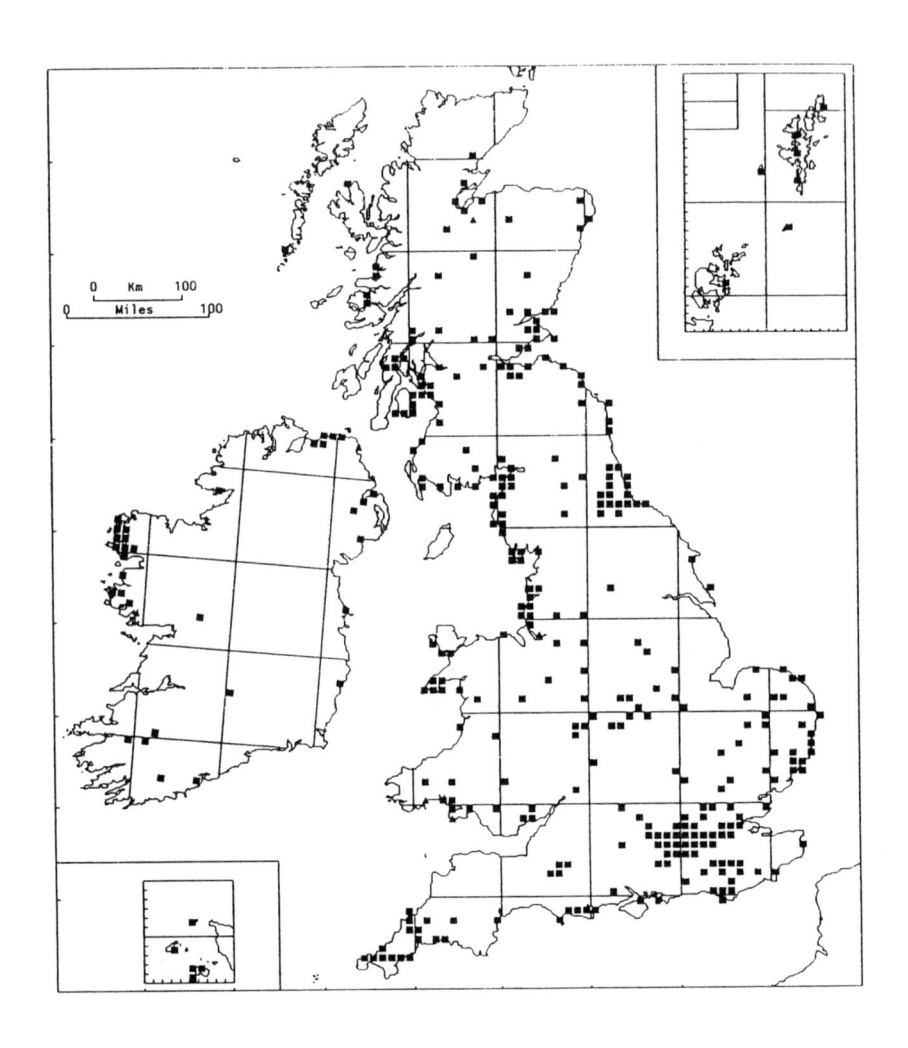

Map 11. **R. stylosa 11**

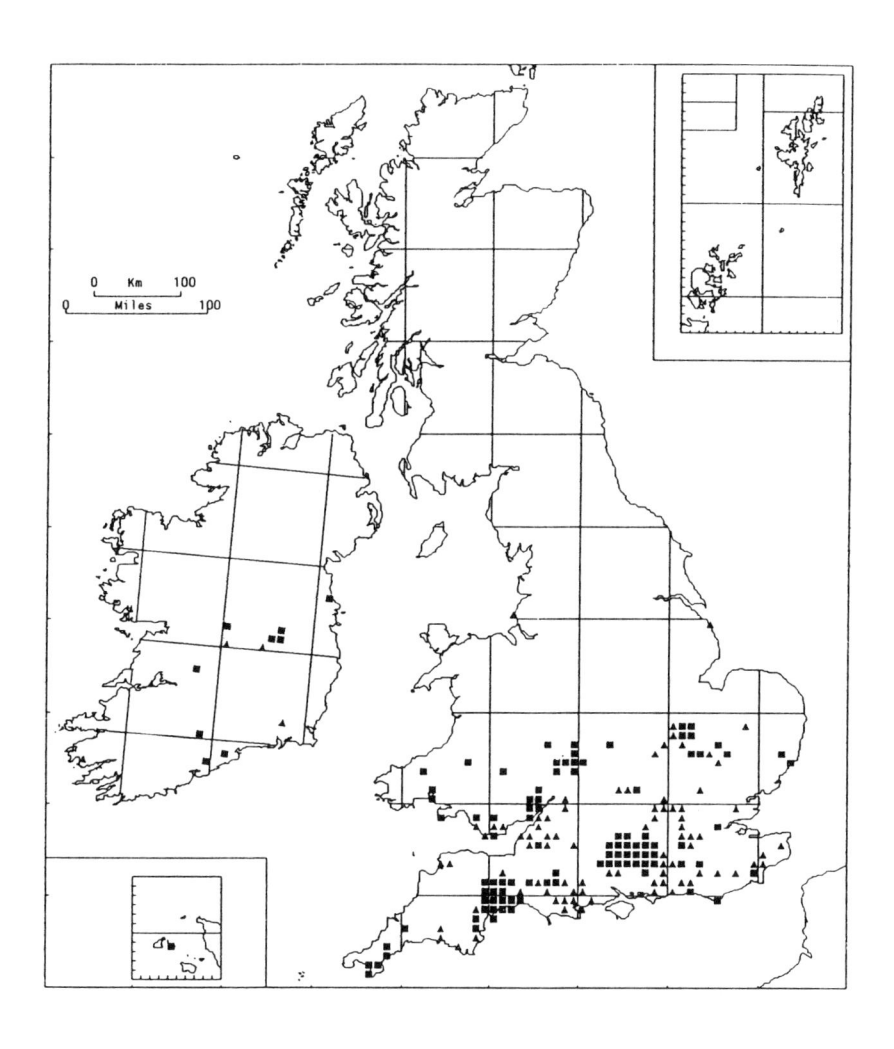

Map 12. **R. × andegavensis**
(**R. canina** × **R. stylosa**) 12 × 11

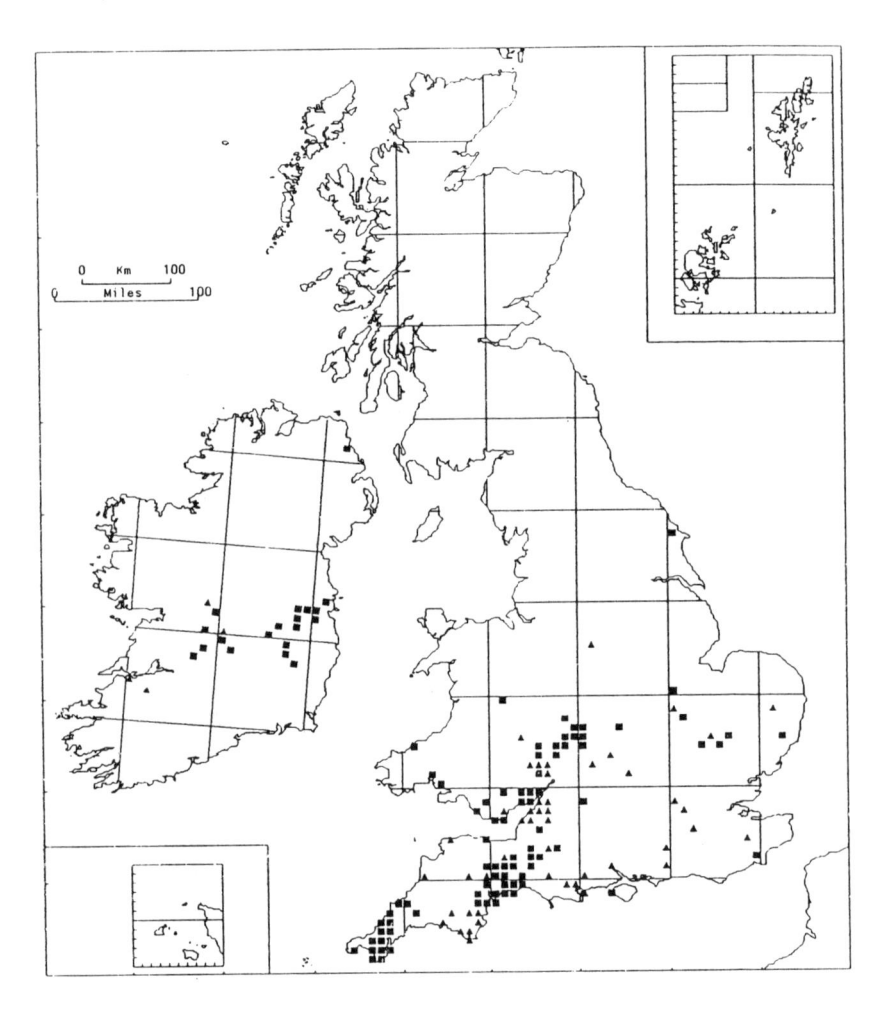

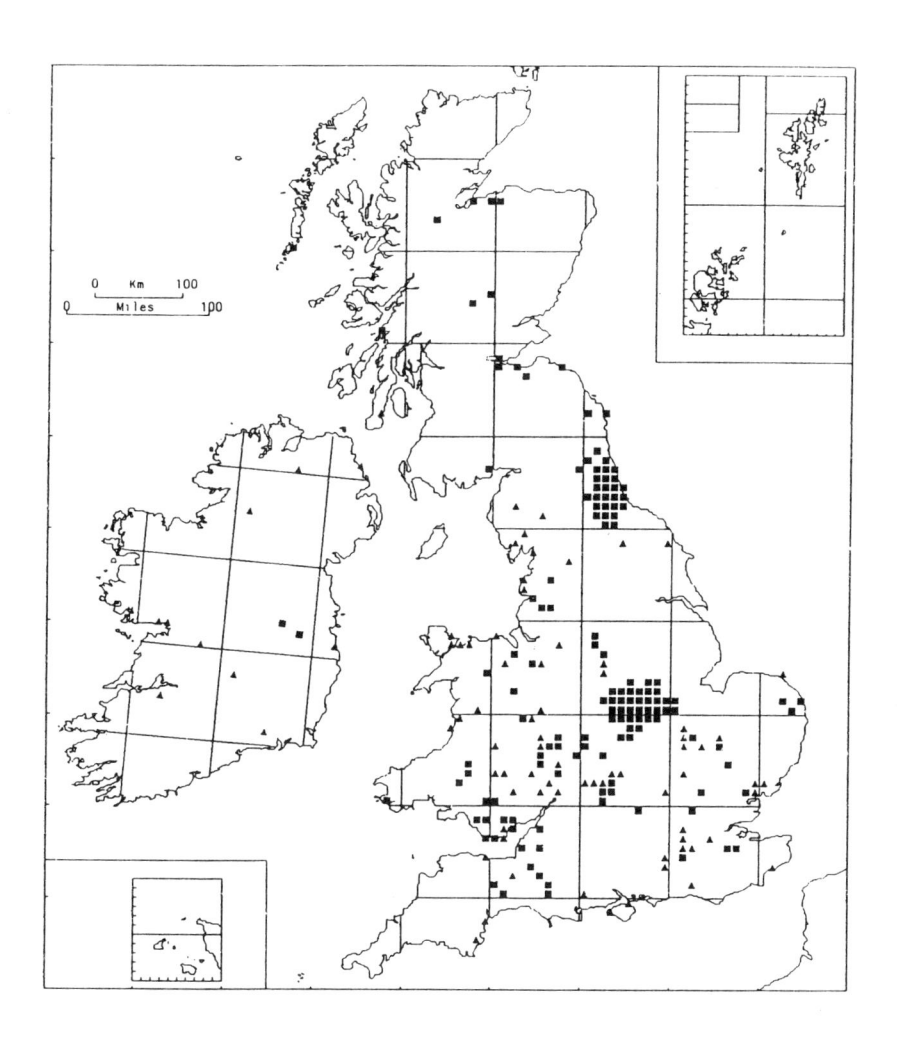

Map 13. **R. canina** Group **Lutetianae 12**

Map 14. **R. canina** Group **Dumales 12**

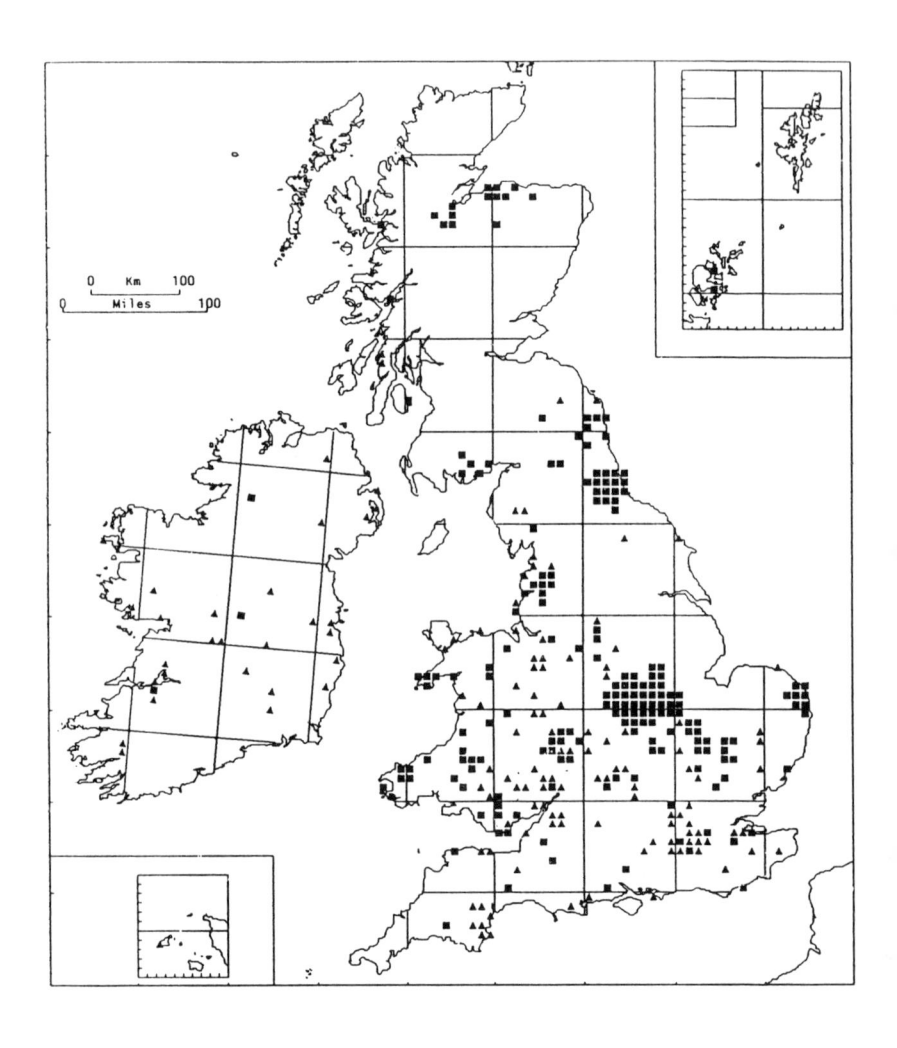

Map 15. **R. canina** Group **Transitoriae 12**

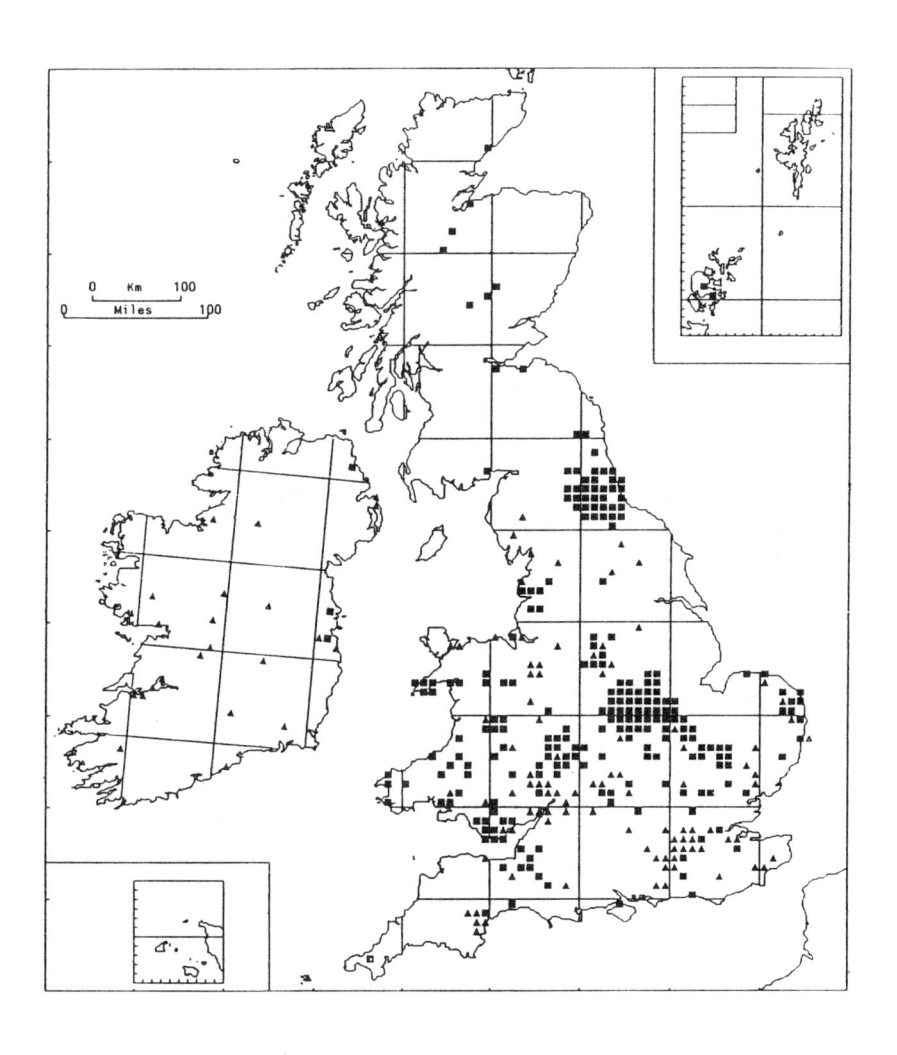

Map 16. **R. canina** Group **Pubescentes 12**

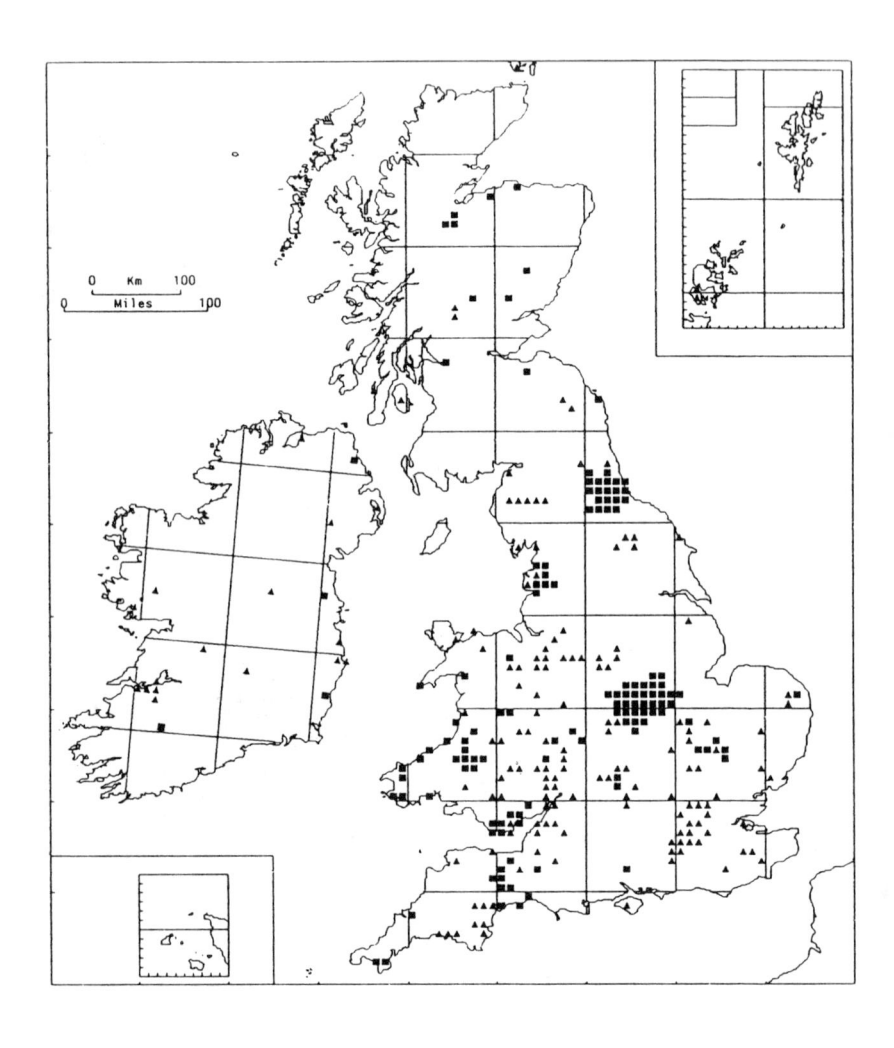

Map 17. **R.** x **dumalis**
   (**R. caesia** subsp. **caesia** x **R. canina**) **13a** x **12**

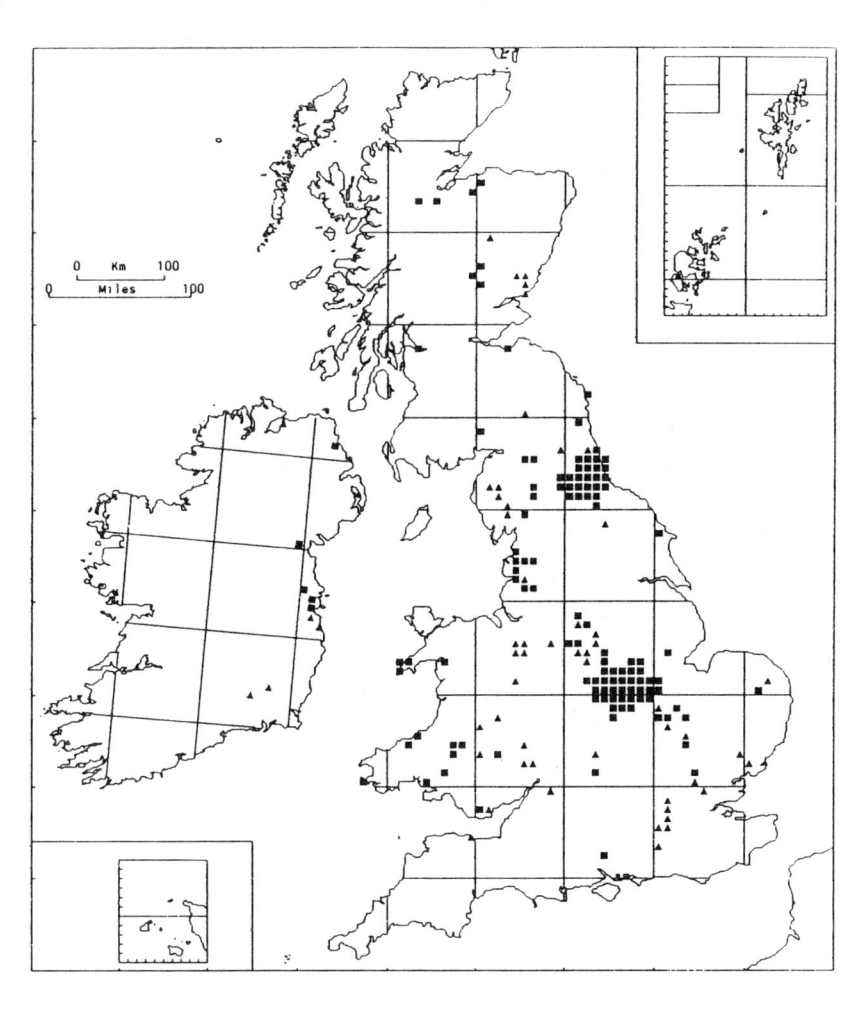

Map 18. **R. × dumalis**
   (**R. caesia** subsp. **glauca** × **R. canina**) **13b × 12**

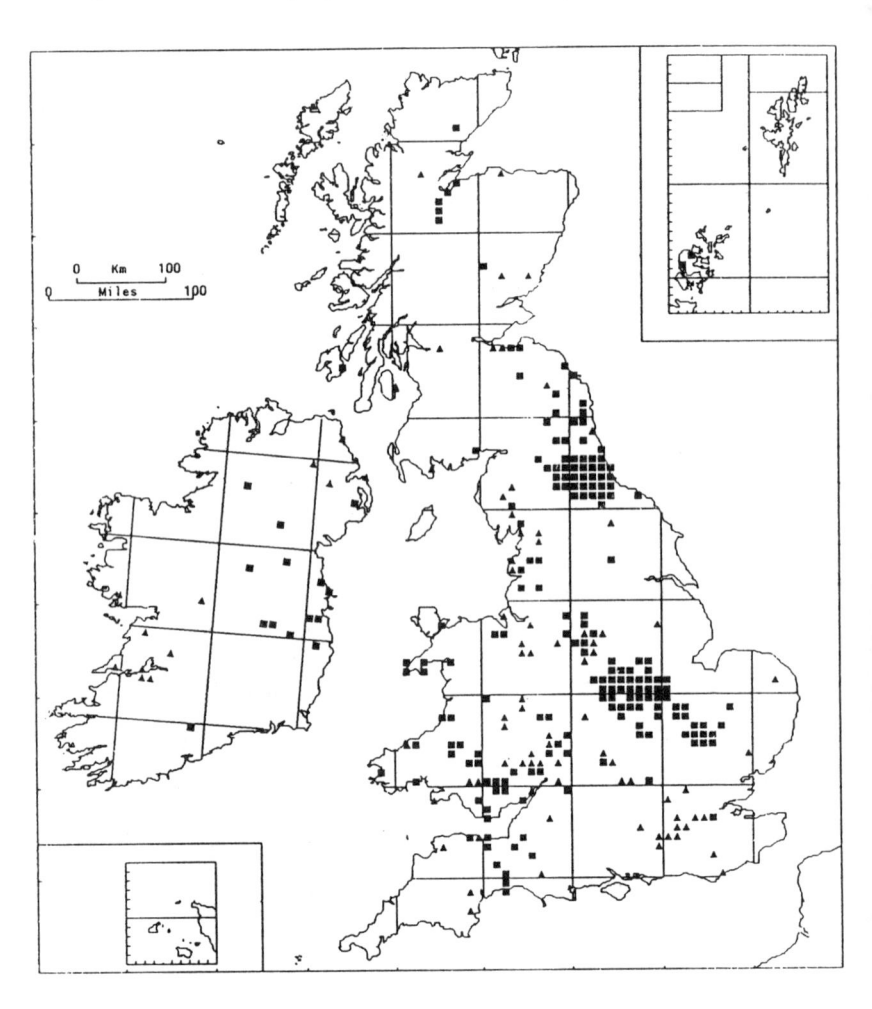

Map 19. **R. × dumetorum**
(**R. canina** × **R. obtusifolia**) **12** × **14**

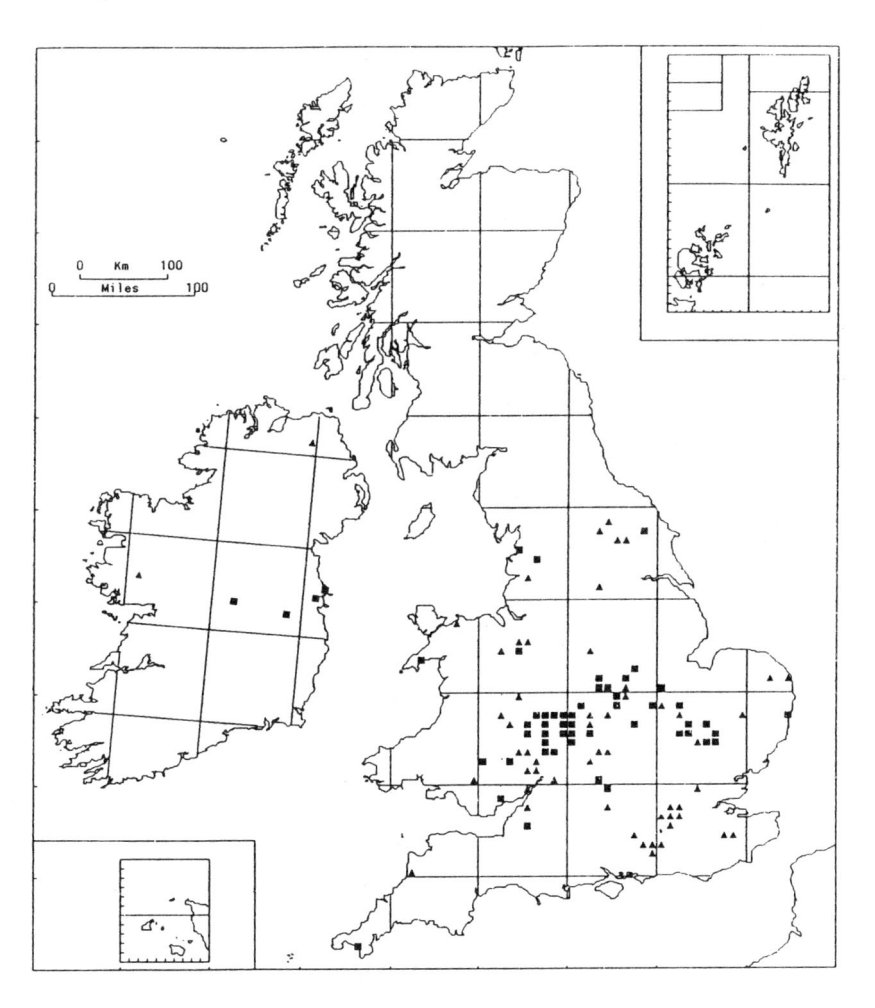

Map 20. **R. × scabriuscula**
(**R. canina** × **R. tomentosa**) 12 × 15

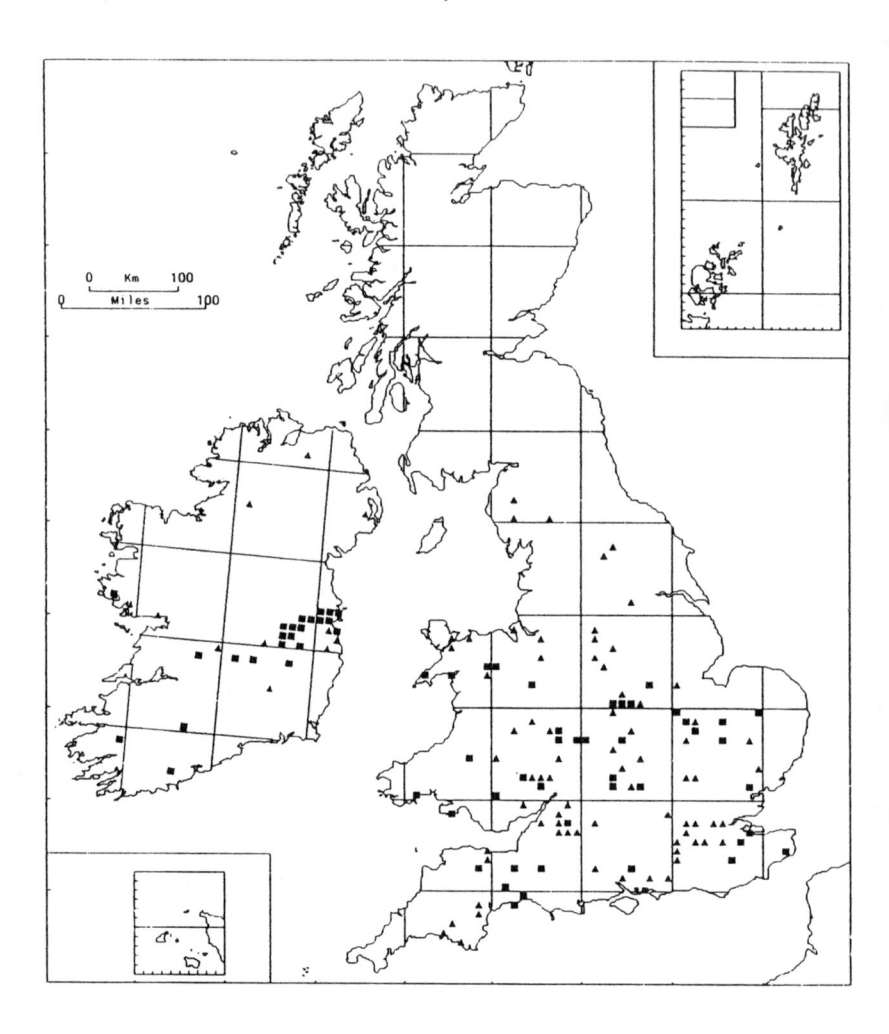

Map 21. **R. × rothschildii**
(**R. canina** × **R. sherardii**) **12 × 16**

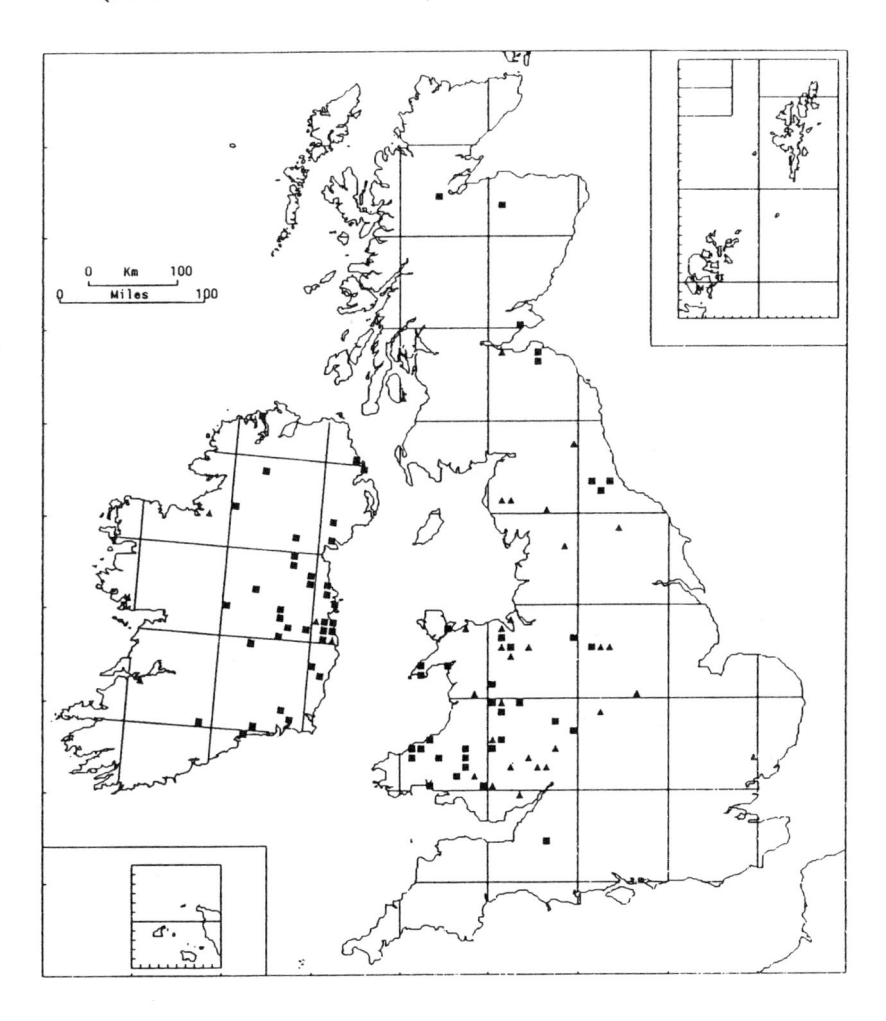

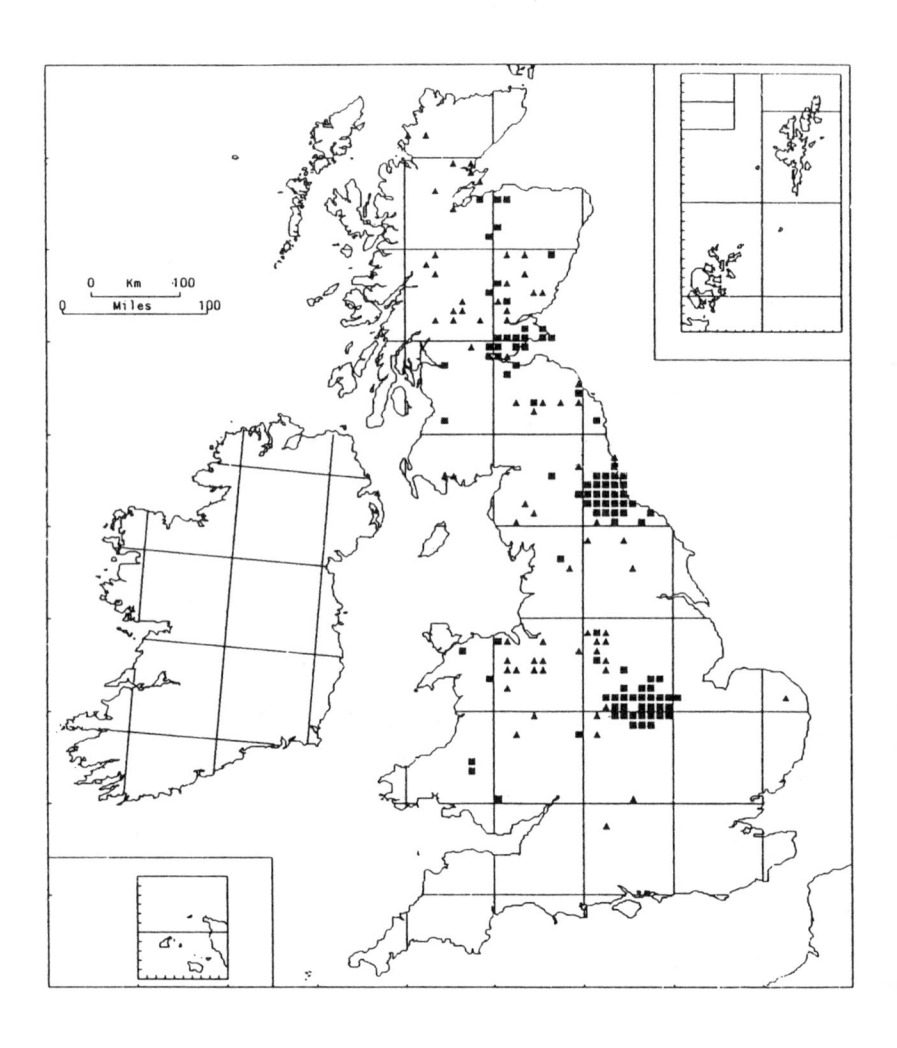

Map 22. **R. caesia** subsp. **caesia 13a**

Map 23. **R. caesia** subsp. **glauca 13b**

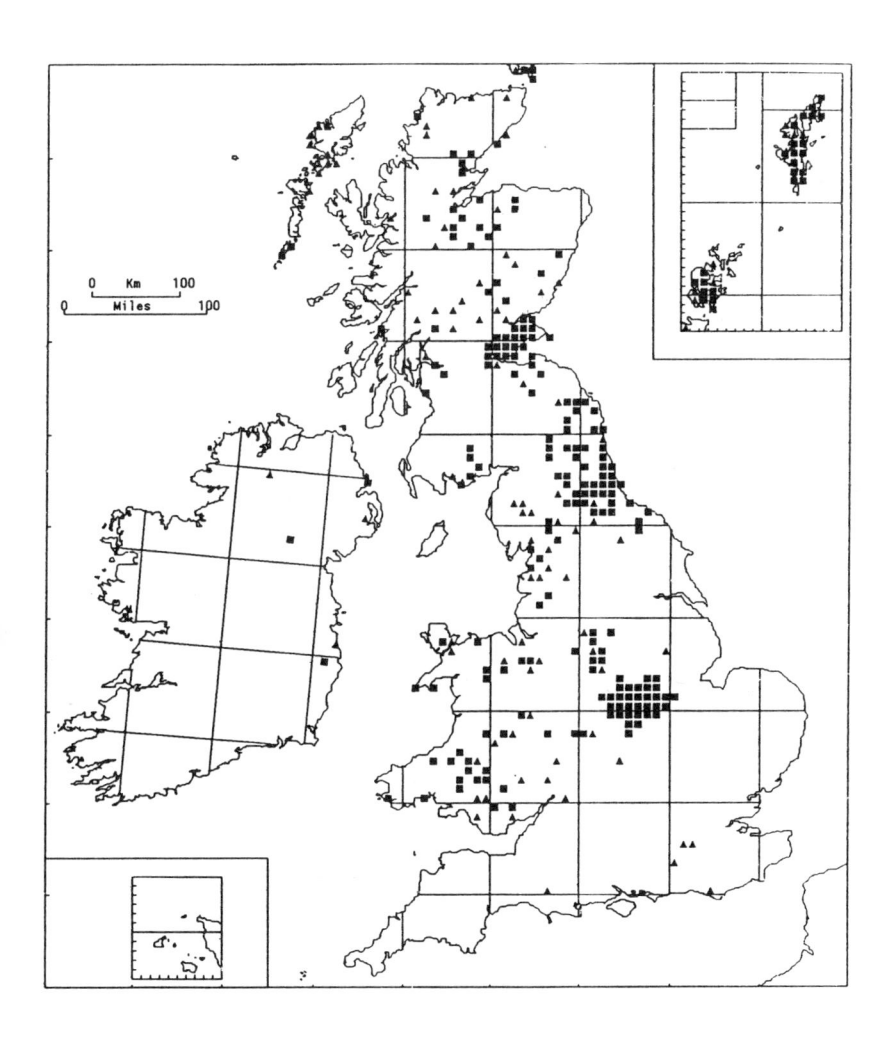

Map 24. **R. caesia** sens. lat. × **R. sherardii** 13 × 16

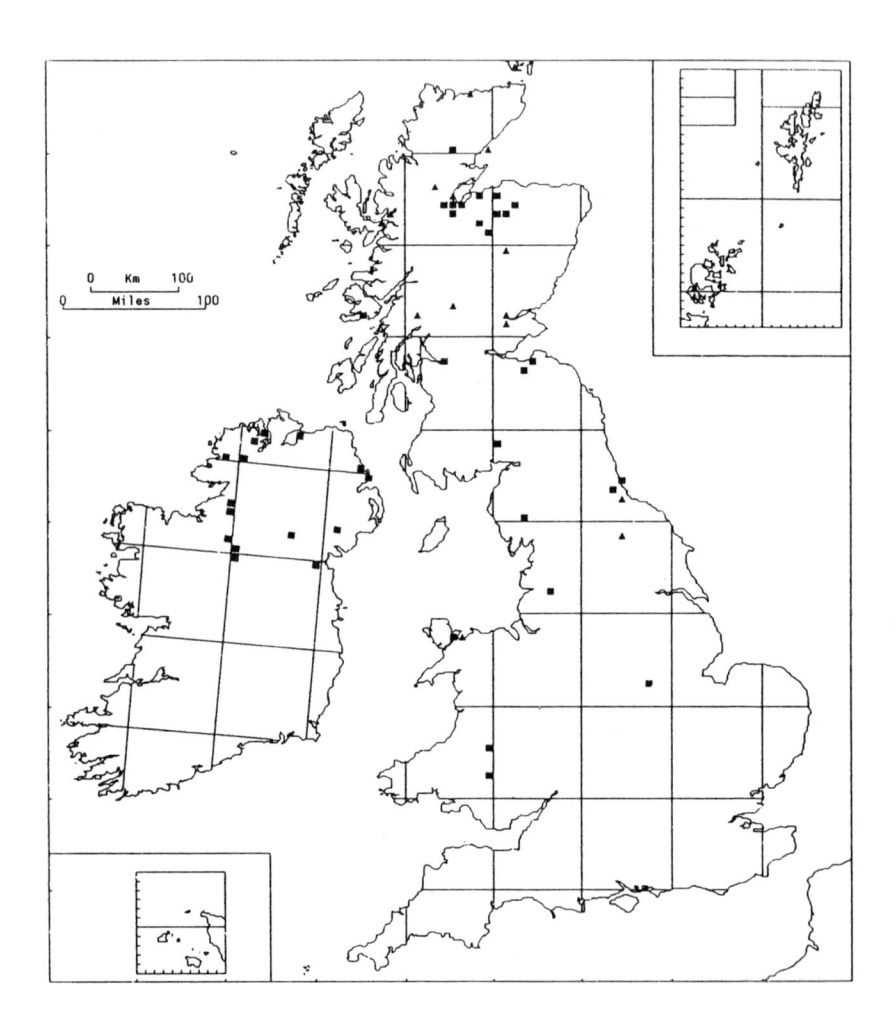

Map 25. **R. obtusifolia 14**

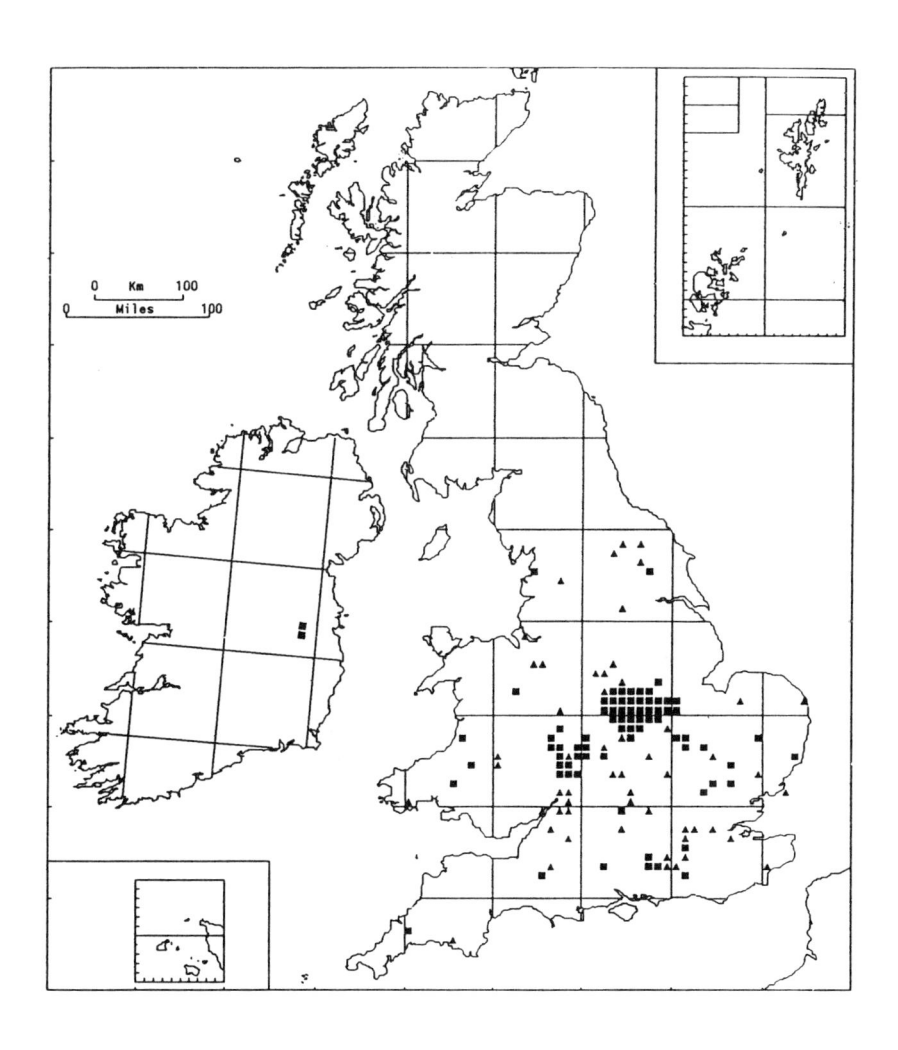

Map 26. **R. tomentosa 15**

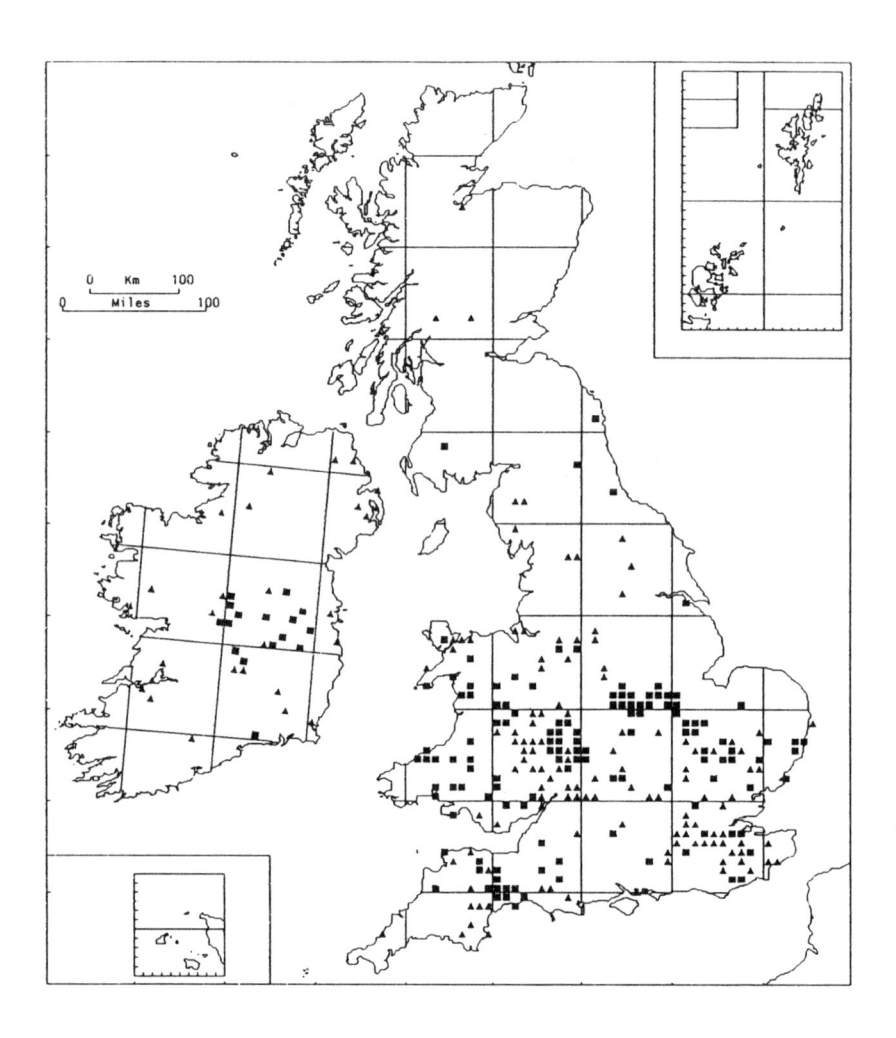

Map 27. **R. sherardii 16**

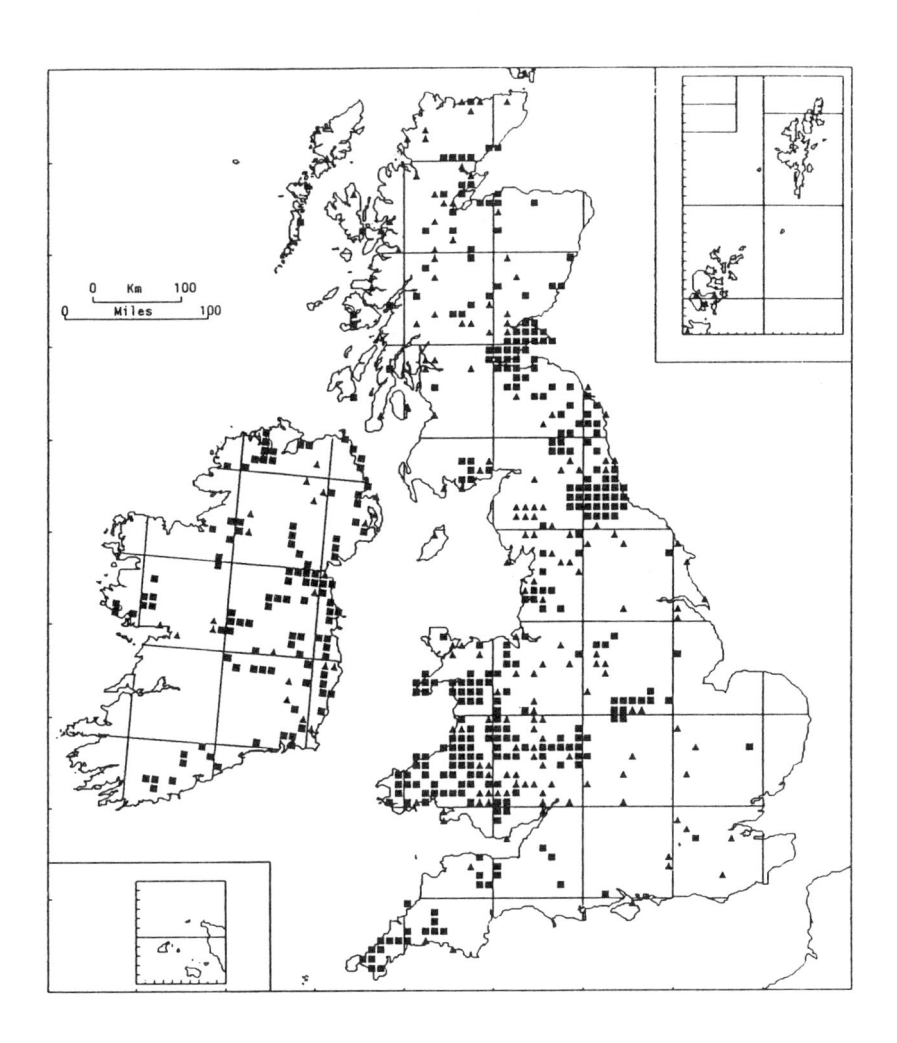

Map 28. **R. x shoolbredii
(R. mollis x R. sherardii) 17 x 16**

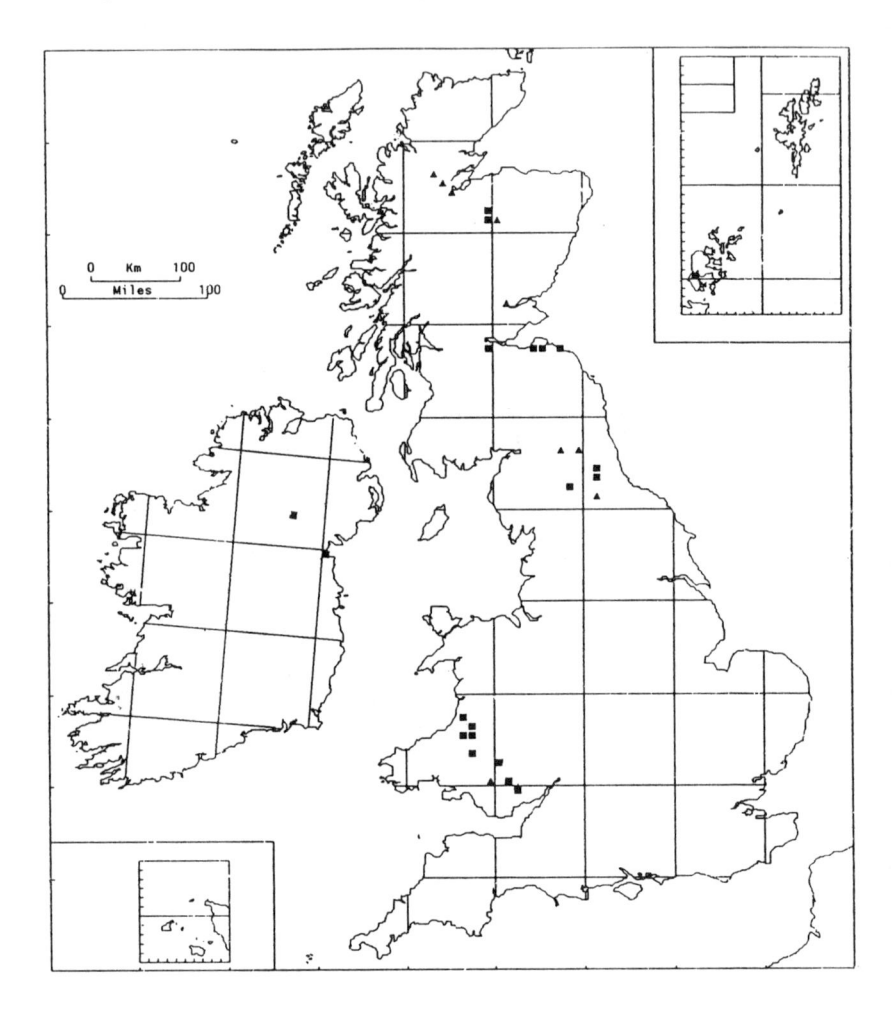

Map 29. **R. mollis 17**

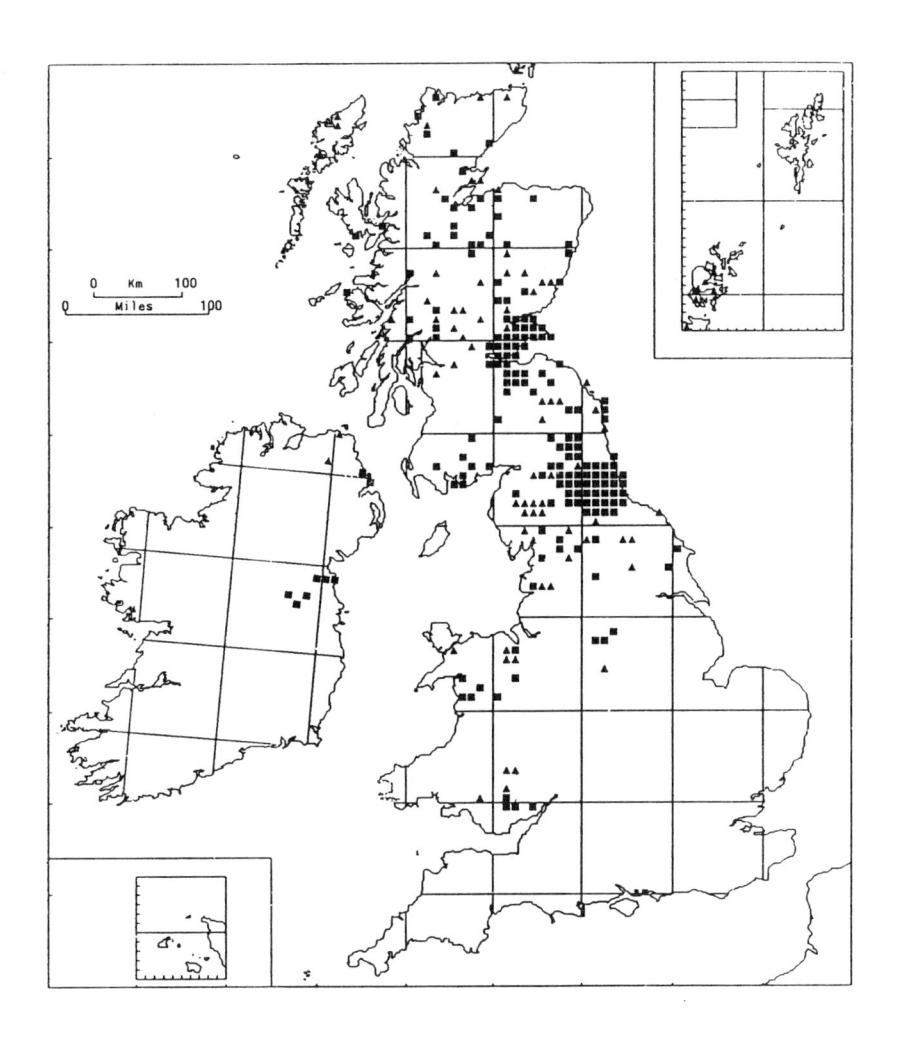

Map 30. **R. rubiginosa 18**

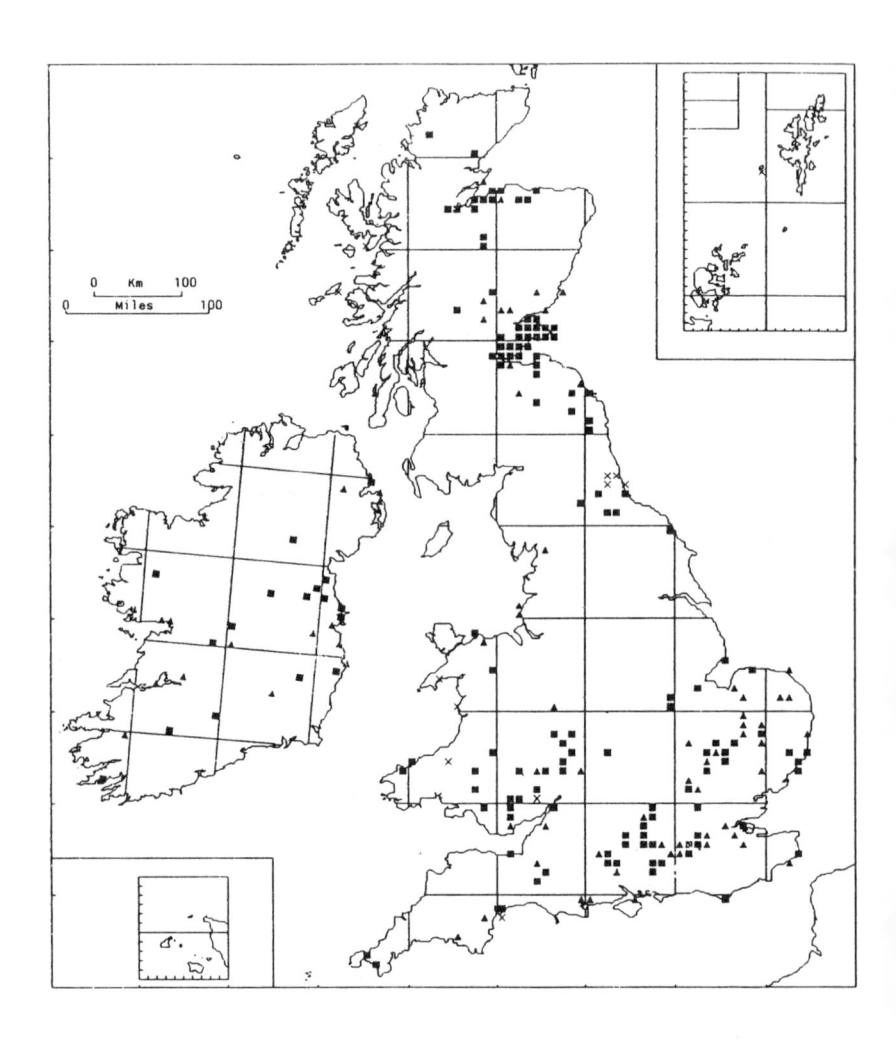

Map 31. **R. micrantha 19**

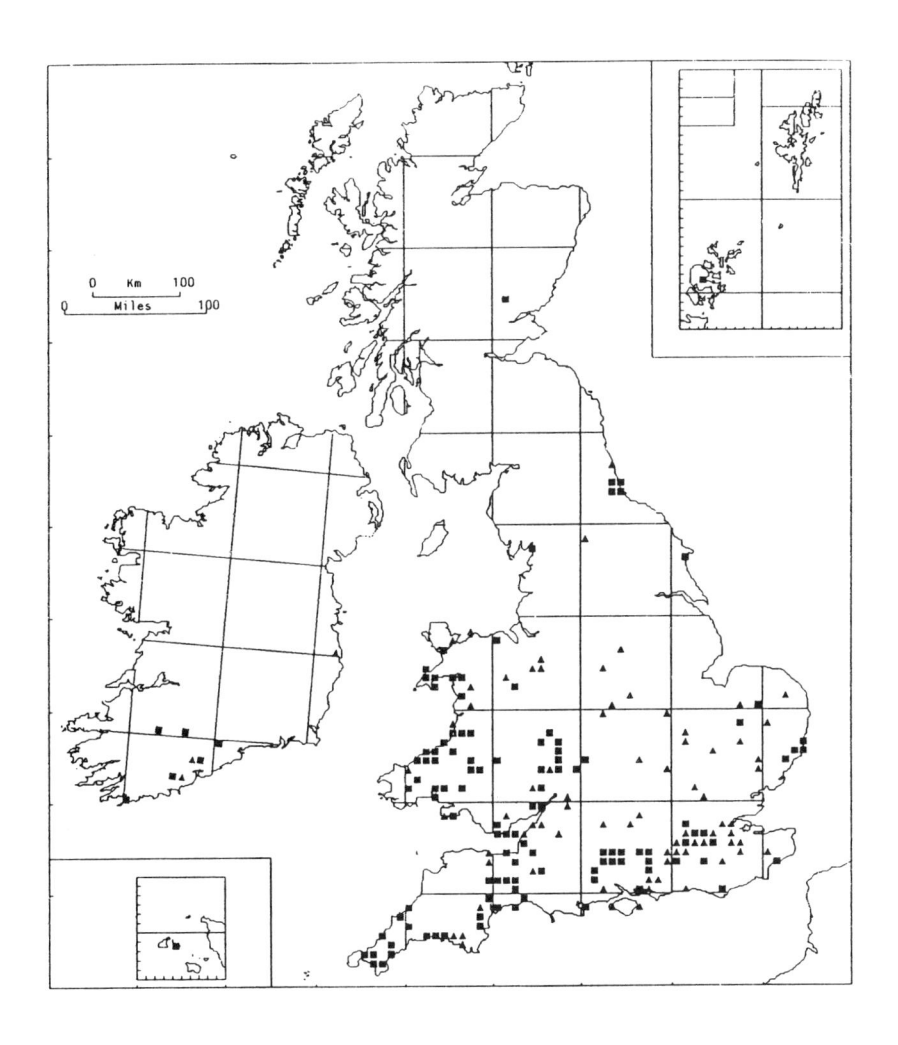

Map 32. **R. agrestis 20**

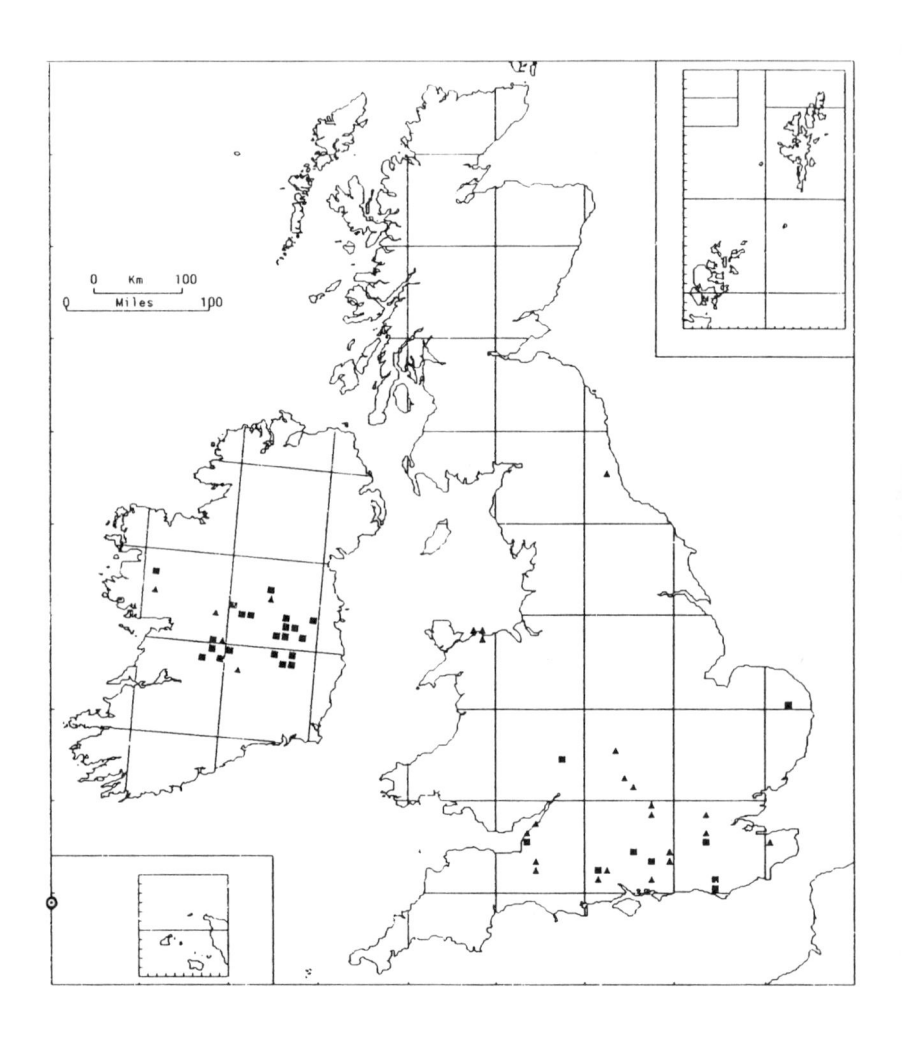

13. Vice-counties of Great Britain and Ireland

England, Wales and Scotland

1. W. Cornwall with Scilly
2. E. Cornwall
3. S. Devon
4. N. Devon
5. S. Somerset
6. N. Somerset
7. N. Wilts.
8. S. Wilts.
9. Dorset
10. Isle of Wight
11. S. Hants.
12. N. Hants.
13. W. Sussex
14. E. Sussex
15. E. Kent
16. W. Kent
17. Surrey
18. S. Essex
19. N. Essex
20. Herts.
21. Middlesex
22. Berks.
23. Oxon
24. Bucks.
25. E. Suffolk
26. W. Suffolk
27. E. Norfolk
28. W. Norfolk
29. Cambs.
30. Beds.
31. Hunts.
32. Northants.
33. E. Gloucs.
34. W. Gloucs.
35. Mons.
36. Herefs.
37. Worcs.
38. Warks.
39. Staffs.
40. Salop
41. Glam.
42. Brecs.
43. Rads.
44. Carms.
45. Pembs.
46. Cards.
47. Monts.
48. Merioneth
49. Caerns.
50. Denbs.
51. Flints.
52. Anglesey
53. S. Lincs.
54. N. Lincs.
55. Leics. with Rutland
56. Notts.
57. Derbys.
58. Cheshire
59. S. Lancs.
60. W. Lancs.
61. S.E. Yorks.
62. N.E. Yorks.
63. S.W. Yorks.
64. Mid-W. Yorks.
65. N.W. Yorks.
66. Co. Durham
67. S. Northumb.
68. Cheviot
69. Westmorland with Furness
70. Cumberland
71. Man
72. Dumfries
73. Kirkudbrights.
74. Wigtowns.
75. Ayrs.
76. Renfrews.
77. Lanarks.
78. Peebless.
79. Selkirks.
80. Roxburghs.
81. Berwicks.
82. E. Lothian
83. Midlothian
84. W. Lothian
85. Fife
86. Stirlings.
87. W. Perth
88. Mid Perth
89. E. Perth
90. Angus
91. Kincardines.
92. S. Aberdeen
93. N. Aberdeen
94. Banffs.
95. Moray
96. Easterness with Nairns.
97. Westerness
98. Main Argyll
99. Dunbarton
100. Clyde Is.
101. Kintyre
102. S. Ebudes
103. Mid Ebudes
104. N. Ebudes
105. W. Ross
106. E. Ross
107. E. Sutherland
108. W. Sutherland
109. Caithness
110. Outer Hebrides
111. Orkney
112. Shetland

Ireland

- H1. S. Kerry
- H2. N. Kerry
- H3. W. Cork
- H4. Mid Cork
- H5. E. Cork
- H6. Co. Waterford
- H7. S. Tipperary
- H8. Co. Limerick
- H9. Co. Clare
- H10. N. Tipperary
- H11. Co. Kilkenny
- H12. Co. Wexford
- H13. Co. Carlow
- H14. Laois
- H15. S.E. Galway
- H16. W. Galway
- H17. N.E. Galway
- H18. Offaly
- H19. Co. Kildare
- H20. Co. Wicklow
- H21. Co. Dublin
- H22. Meath
- H23. Westmeath
- H24. Co. Longford
- H25. Co. Roscommon
- H26. E. Mayo
- H27. W. Mayo
- H28. Co. Sligo
- H29. Co. Leitrim
- H30. Co. Cavan
- H31. Co. Louth
- H32. Co. Monaghan
- H33. Fermanagh
- H34. E. Donegal
- H35. W. Donegal
- H36. Tyrone
- H37. Co. Armagh
- H38. Co. Down
- H39. Co. Antrim
- H40. Co. Londonderry

Channel Isles CI

Vice-counties

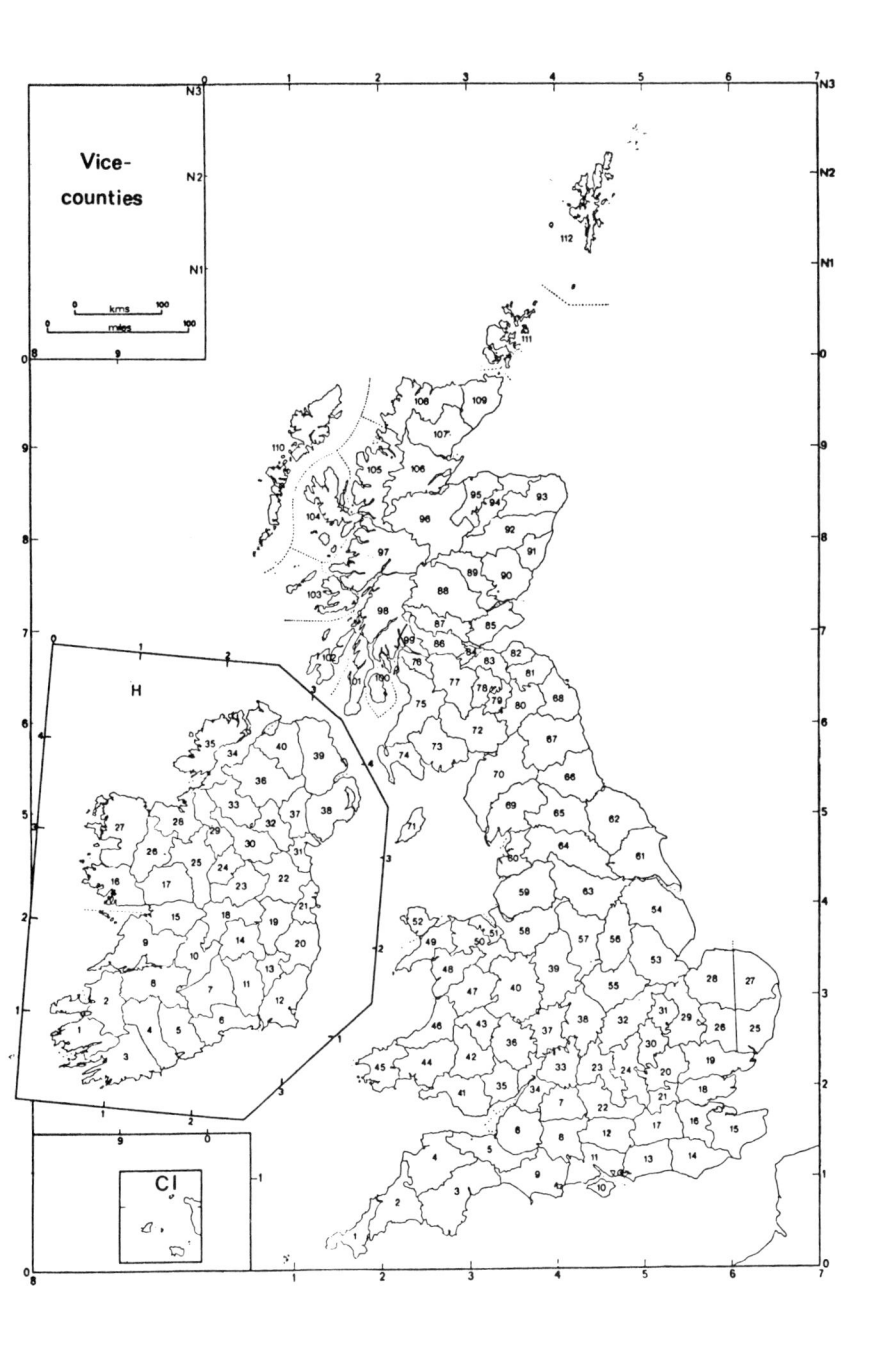

14. Glossary

The terms are defined in the sense in which they are used in this Handbook, and are not necessarily used in the same sense in other works.

Achene: a single-seeded nut-like fruit; the true fruit of roses (see hip).
Acicle: a small, very slender, nearly straight prickle, tapering to a point but not broadened at the base, sometimes gland-tipped.
Aciculate: bearing acicles.
Acuminate: with a long, fine point.
Acute: sharply and shortly pointed, the angle made by the point being less than 90 degrees.
Adnate: attached (to another structure) along the greater part of its length.
Anthesis: the time of expansion of a flower when pollination takes place.
Arcuate: (of prickles), curved like a bow, with the upper and lower margins forming circular arcs of large diameter.
Armature: the prickles and acicles of roses.
Axil: the upper angle between a leaf and the stem, or between a leaflet and the rachis.
Biserrate: describes the margin of a leaflet which has large, sharp teeth each with one smaller one arising on its lower side.
Blade: the expanded part of a leaf.
Bract: a modified leaf in the axil of which a flower bud develops.
Caesious: lavender-blue-grey.
Calyx: the outer (or lowest) part of the flower, consisting (in *Rosa*) of five sepals which form the cover of the flower-bud.
Coriaceous: leathery.
Crenate-serrate: describes teeth which are rounded over most of their edges but end in a sharp point.
Cuneate: wedge-shaped, i.e. with straight sides set at an acute angle.
Declining: (of prickles) straight or nearly so but directed sensibly downwards.
Deltate: more or less equilaterally triangular.
Denticle: a smaller secondary tooth arising from the side of a larger primary tooth.
Depressed-globose: globose, but depressed at the poles.
Disc: the central portion of the top of the hip within the ring of stamens.
Eglandular: without glands.
Ellipsoid: solid form of elliptic.

Elliptic: broadest at the middle with margins symmetrically curved; length c. 1.75 × breadth.
Entire: margins even, not toothed or lobed or cut.
Erect: (of sepals) in a more or less vertical position in relation to the disc; (of stems) in a more or less vertical position in relation to the substrate.
Exserted: (of styles) protruding from the orifice of the hip so that parts of the styles are visible above the level of the disc.
Flexuous: describes young stems (usually short) which bend slightly under the influence of gravity.
Fusiform: (of hips) broadly spindle-shaped or narrowly ellipsoid.
Glabrous: lacking hairs.
Gland: small knob-like structure, sessile or stalked, which is presumed to be secretory.
Glandular-hispid: (of pedicels) clothed with small, stalked or almost sessile glands.
Glandular-setose: (of hips) bristly with often poorly developed glands.
Glaucous: bluish-grey.
Globose: more or less spherical.
Gynoecium: the whole female reproductive part of the flower.
Hip: the false fruit of a rose, consisting of a fleshy concave receptacle surrounding the true fruits or achenes.
Hispid: bearing erect, straight, harshly stiff hairs.
Inflorescence: collective term for all the flowering part of a branch arising from one main axis.
Internode: the length of stem between two nodes (q.v.).
Irregularly uniserrate: describes a leaflet which is uniserrate except for a few smaller teeth spaced at irregular intervals between the larger ones.
Laciniate: cut into fine, deep, irregular divisions or lobes.
Lanceolate: shaped like the head of a lance, broadest below the middle and more than three times as long as wide.
Leaflet: one of the elements of a compound leaf, arising from the petiole or rachis, not from the stem.
Midrib: the central, main vein of a leaflet.
Multiserrate: describes a leaflet with sharp primary teeth from one or both sides of each of which arise two or more small secondary teeth.
Node: the point on a stem from which a leaf or leaves arise.
Ob-: inverted e.g. obovoid.
Obovoid: shaped like an egg with the small end at the bottom.

Ovate: egg-shaped in outline, with rounded margins, broadest below the middle, and less than twice as long as wide.
Ovoid: shaped like an egg with the small end at the top; the solid form of ovate.
Patent: (of prickles) at right angles to the stem.
Pedicel: the stalk of a single flower.
Persistent: (of sepals) remaining attached to the hip whilst the fruit ripens.
Petiole: the stalk of a leaf, below the lowest pair of leaflets.
Pilose: covered with soft, very long, rather straight hairs, not dense, but somewhat shaggy.
Pinnate: of a leaf, with leaflets arranged in opposite pairs (and, in *Rosa*, with a single terminal leaflet); of a sepal, with lateral lobes arranged in opposite pairs.
Prickle: a spine-like process with a broadened base and a sharp point.
Pricklet: small prickle similar in shape to the main prickles (distinct from an acicle in that the latter is usually more slender and is straighter).
Pubescent: a general term for hairy as distinct from glabrous.
Rachis: the axis of the leaf between the lowest pair of leaflets and the terminal leaflet.
Receptacle: the modified end of the flower stalk bearing the flower or fruits.
Reflexed: (of sepals) turned more or less vertically downwards.
Rugose: with a ridged irregular surface.
Scabrid: covered with scattered, stiff, inclined hairs (often with bulbous bases) which make the surface feel rough to the touch.
Serrate: with sharp, straight-sided teeth like a saw.
Sessile: not stalked.
Simple: undivided; not branched or compound.
Spreading: (of sepals) arranged more or less horizontally in relation to the hip.
Spreading-erect: (of sepals) arranged more or less at 45 degrees above the disc of the hip.
Stigma: the pollen-receptive end of the style.
Stipitate: with a short, slender stalk.
Stipule: one of a pair of appendages at the base of the petiole.
Style: the part of the gynoecium connecting the ovary with the stigma. In a rose hip each achene has its own filamentous style, all of which come together in a bundle (fused together in some species to form a stylar column) at the orifice of the hip.
Subulate: awl-shaped.

Sweet-briar-type: (of glands), brownish or translucent, viscid, c. 100-120µm in diameter, with an odour reminiscent of ripe apples when crushed.
Terete: with a circular cross-section, without grooves or protuberances.
Tomentose: densely covered with short, cottony hairs.
Turbinate: (of hips) top-shaped.
Uniserrate: describes a leaflet with teeth which are sharp and all more or less of the same size.
Urceolate: (of hips) roughly urn-shaped, somewhat constricted below the disc.
Villous: covered with moderately dense, long, soft, often curly hairs, which tend to be erect though they are not necessarily straight.
Viscid: moistly sticky, like treacle.

15. Select bibliography and references

ALLEN, E.F. 1973. A simplified Rose classification for gardeners. *The Rose Annual* **1973**:133-139.

ALMQUIST, S.I. 1918. *Rosa* L. In Lindman, C.A.M. *Svensk Fanerogamflora*: 335-379. Stockholm.

ALMQUIST, S.I. 1919. *Sveriges Rosae*. Stockholm.

ARNAIZ, C., GEHU, J.-M. & GEHU-FRANCK, J. 1980. Apport à la connaissance des espèces du genre *Rosa* dans la région Nord-Pas-de-Calais (France). *Bulletin de la Société botanique de N. France* **33**(3-4):65-83.

BAKER, J.G. 1864a. Review of the British Roses, especially those of the North of England (parts I-V). *The Naturalist* **1**:14-24,33-38,60-67,93-103,141-144.

BAKER, J.G. 1864b. *Rosa alpina* L. in Britain. *The Naturalist* **1**:184-186.

BAKER, J.G. 1865. On *Rosa collina* Jacq. as a British plant. *Journal of Botany* **3**:82-84.

BAKER, J.G. 1867. *Rosa inodora* Fries. *Journal of Botany* **5**:65-66.

BAKER, J.G. 1869. A monograph of the British Roses. *Journal of the Linnean Society. Botany* **11**:197-243.

BAKER, J.G. 1870a. British Roses. *Journal of Botany* **8**:24-25.

BAKER, J.G. 1870b. On *Rosa sepium* Thuill., and other new or little-known forms of British Roses. *Journal of Botany* **8**:77-80.

BAKER, J.G. 1870c. *Rosa Sabini* in France. *Journal of Botany* **8**:161-162.

BAKER, J.G. 1873. On *Rosa apennina* Woods. *Journal of Botany* **11**:35-36.

BAKER, J.G. 1875. On the botanical origin of attar of Roses. *Journal of Botany* **13**:8.

BAKER, J.G. 1878. *Rosa hibernica* Sm. var. *Grovesii* Baker. *Journal of Botany* **16**:183-184

BAKER, J.G. 1885. A classification of garden Roses. *Journal of Botany* **23**:281-286.

BAKER, J.G. 1892. On a new form of *Rosa tomentosa* Woods. *Journal of Botany* **30**:341-342.

BAKER, J.G. 1905. A revised classification of British Roses. *Journal of the Linnean Society. Botany* **37**:70-79.

BALLANTYNE, G. & GRAHAM, G.G. 1979. Roses and Brambles, Kindrogan. *Botanical Society of Edinburgh News*, **27**:16-19.

BARCLAY, W. 1899. Further notes on Scottish Roses. *Annals of Scottish natural History* **1899**:172-179.

BARCLAY, W. 1908. The genus *Rosa* in the 'London Catalogue,' Ed. 10. *Journal of Botany* **46**:278-280,356-358.

BARCLAY, W. 1910a. Note on *Rosa*. *Journal of Botany* **48**:187.

BARCLAY, W. 1910b. Perthshire Roses. *Proceedings of the Perthshire Society of Natural Science* **5**(2):66-74.

BARCLAY, W. 1910c. Note upon *Rosa*. *Journal of Botany* **48**:205.

BARCLAY, W. 1910d. *Rosa pimpinellifolia* L. × *Rosa rubiginosa* L. *Journal of Botany* **48**:260.

BARCLAY, W. 1910e. A new variety of *Rosa Hibernica*. *Journal of Botany* **48**:332-333.

BARCLAY, W. 1911. Our native hybrid Roses. *Proceedings of the Perthshire Society of natural Science* **5**(3):112-117.

BARCLAY, W. 1915. Notes on Roses, part I. *Proceedings of the Perthshire Society of natural Science* **6**(2):82-89.

BARCLAY, W. 1916. Notes on Roses, part II. *Proceedings of the Perthshire Society of natural Science* **6**(3):115-124.

BARRINGTON, R.M. 1876. *Rosa Britannica*. *Journal of Botany* **14**:270.

BASTARD, T. 1809. *Essai sur la Flore du Département de Maine et Loire* 185-189. Angers.

BASTARD, T. 1812. *Essai sur la Flore du Département de Maine et Loire, Supplément*:29-33. Angers.

BEALES, P. 1985. *Classic Roses*. London.

BECHSTEIN, J.M. 1810. *Forstbotanik*.

BESSER, W.S.J.G. von 1822. *Enumeratio Plantarum*:19-21,59-69. Vilnae.

BEST, G.N. 1887. Remarks on the Group *Carolinae* of the genus *Rosa*. 1. *Bulletin of the Torrey botanical Club* **14**(2):253-256.

BEST, G.N. 1889a. Remarks on the Group *Carolinae* of the genus *Rosa*. 2. *Bulletin of the Torrey botanical Club* **16**(6):161-165.

BEST, G.N. 1889b. North American Roses: remarks on characters with classification. *Journal of the Trenton natural History Society* **2**(1):1-8.

BEST, G.N. 1890. Remarks on the Group *Cinnamomeae* of the North American Roses. *Bulletin of the Torrey botanical Club* **17**(6):141-149.

BISHOP, E.B. 1931. New British Roses from Northumberland. – J.W.Heslop Harrison [Abstract]. *Journal of Botany* **69**:15-16.

BISHOP, E.B. 1933. A new variety of *Rosa micrantha* Sm. *Report of the botanical Society and Exchange Club of the British Isles* **10**:468-471.

BLACKBURN, K.B. 1925. Chromosomes and classification in the genus *Rosa*. *American Naturalist* **59**:200-205.

BLACKBURN, K.B. 1927a. Chromosomes and their relation to Rose problems. *American Rose Annual* **1927**:54-58.

BLACKBURN, K.B. 1927b. A new rose from Northumberland. *University of Durham philosophical Society Proceedings* **8**:101-103, t. opp. p. 100.

BLACKBURN, K.B. 1930. *Rosa glauca* Vill. var. *berniciensis* Blackburn. *Journal of Botany* **68**:121.

BLACKBURN, K.B. 1949. Chromosomes and classification in the genus *Rosa*. In Wilmott, A.J. (ed.) *British flowering Plants and modern systematic Methods*:53-57. London.

BLACKBURN, K.B. & HARRISON, J.W.H. 1921. The status of the British Rose forms as determined by their cytological behaviour. *Annals of Botany* **35**:159-188, pl. 9 & 10.

BLACKBURN, K.B. & HARRISON, J.W.H. 1924. Genetical & cytological studies in hybrid roses. 1. The origin of a fertile hexaploid form in the *Pimpinellifoliae – Villosae* crosses. *British Journal of experimental Biology* **1**(4): 557-570, pl.1 & 2.

BOLTON, E. 1934. A Durham hybrid between *Rosa pimpinellifolia* and *Rosa mollis*. *The Vasculum* **20**:90.

BOLTON, E. 1940. Interesting facts concerning our wild Roses. *The Vasculum* **26**:4-6.

BOLTON, E. 1942. Observations on *Rosa sherardii* var. *glabrata* (Fries) Boulenger. *Journal of Botany* **80**:149-152.

BORBAS, V. 1880. *Primitiae Monographiae Rosarum Imperii hungarici*. Budapest.

BOREAU, A. 1840. *Flore du Centre de la France*:171-183. Paris.

BORKHAUSEN, M.B. 1790. *Versuch einer forstbotanischen Beschreibung*: 296-334. Frankfurt am Main.

BOSWELL-SYME, J.T.B. 1864. *English Botany*, Ed. 3, **3**:202-233, tt.461-476. London.

BOULENGER, G.A. 1920a. Some Roses from Dorsetshire. *Journal of Botany* **58**:16-21

BOULENGER, G.A. 1920b. On *Rosa Britannica* Déséglise. *Journal of Botany* **58**:185-187.

BOULENGER, G.A. 1924-25. Les Roses d'Europe de l'herbier Crépin. *Bulletin du Jardin botanique de l'État à Bruxelles* **10**(1):1-192;**10**(2):193-417.

BOULENGER, G.A. 1931-32. Les Roses d'Europe de l'herbier Crépin. *Bulletin du Jardin botanique de l'État à Bruxelles* **12**(1):1-192;**12**(2/3):193-242.

BOULENGER, G.A. 1927. Sur le *Rosa Dumalis* de Bechstein. *Bulletin de la Société royale de Botanique de Belgique* **59**(2):113-115.

BOULENGER, G.A. 1937. Introduction à l'étude du Genre *Rosa*. *Bulletin du Jardin Botanique de l'État à Bruxelles* **14**(3):241-273.

BOULENGER, G.A. 1936. Sur l'allure des sépales après l'anthèse dans le genre *Rosa*. *Bulletin de la Société royale de Botanique de Belgique* **69**(1):60-63.

BRIGGS, T.R.A. 1870. *Rosa saxatilis* Bor. *Report of the botanical Exchange Club* **1869**:16.

BRIGGS, T.R.A. 1877. On the Roses of the neighbourhood of Plymouth. *Journal of Botany* **15**:315-316.

BRITTEN, J. 1870. *Rosa micrantha*. New to Durham [Berwickshire]. *Journal of Botany* **8**:19.

BRITTEN, J. 1907. Note on *Rosa hibernica*. *Journal of Botany* **45**:304-305.

BURNAT, É. & GREMLI, A. 1879. *Les Roses des Alpes Maritimes*. Genève & Bâle.

BURNAT, É. & GREMLI, A. 1882-83. *Supplément à la Monographie des Roses des Alpes Maritimes*. Genève & Bâle.

BUTCHER, R.W. 1961. *A new illustrated British Flora* **1**:695-707. London.

CHANDLER, J.H. & GRAHAM, G.G. 1973. A Lincolnshire Rose problem. *The Naturalist* **1977**:23-24.

CHRIST, K.H.H. 1873. *Die Rosen der Schweiz*. Basel.

CHRIST, K.H.H. 1875a. Neue und bemerkenswerthe Rosenformen. *Flora* **58**(18):273-281; (19):289-297.

CHRIST, K.H.H. 1875b. What is *Rosa hibernica* of Smith? *Journal of Botany* **13**:100-102

CHRIST, K.H.H. 1875c. *Rosa sclerophylla* Scheutz, a new British Rose. *Journal of Botany* **13**:102-103.

CHRIST, K.H.H. 1876. Les Roses des Alpes Maritimes. *Journal of Botany* **14**:137-142,170-172.

CHRIST, K.H.H. 1884. Allgemeine Ergebnisse aus der systematischen Arbeit am Genus *Rosa*. *Botanisches Centralblatt* **18**:310-318,343-350,372-382,385-399.

COGGIATTI, S. 1987. *The Macdonald Encyclopedia of Roses*. London & Sydney.

COLE, R.D. 1917. Imperfection of pollen and mutability in the genus *Rosa*. *Botanical Gazette* **63**:110-123.

CORSTORPHINE, M. 1932. Roses in Angus. *Report of the botanical Society and Exchange Club of the British Isles* **9**:695-709.

CRÉPIN, F. 1860. *Manuel de la Flore de Belgique*:51-52. Brussels.

CRÉPIN, F. 1866. Études sur les Roses. *Bulletin de la Société royale de Botanique de Belgique* **5**:13-27.

CRÉPIN, F. 1870. How to gather Roses. *Journal of Botany* **8**:25-27.

CRÉPIN, F. 1872. *Primitiae monographiae Rosarum*. Plant Monograph Reprints, Cramer, J. & Swan, H.K. (eds.) Brussels.

CRÉPIN, F. 1889. Sketch of a new classification of Roses. *Journal of the Royal Horticultural Society* **11**:217-228.

CRÉPIN, F. 1890. Recherches sur l'état du développement des grains de pollen dans diverses espèces du genre *Rosa*. *Bulletin de la Société royale de Botanique de Belgique* **28**(2):114-125.

CRÉPIN, F. 1892. Tableau analytique des Roses Européennes. *Bulletin de la Société royale de Botanique de Belgique* **31**(2):66-92.

CRÉPIN, F. 1893. L'obsession de l'individu dans l'étude des Roses. *Bulletin de la Société royale de Botanique de Belgique* **32**(2,1):52-55.

CRÉPIN, F. 1894a. *Rosae hybridae*. Études sur les Roses hybrides. *Bulletin de la Société royale de Botanique de Belgique* **33**:7-149.

CRÉPIN, F. 1894b. Sur la nécessité d'une nouvelle monographie des Roses de l'Angleterre. *Bulletin de la Société royale de Botanique de Belgique* **33**(2):14-25.

CRÉPIN, F. 1895. On the necessity for a new monograph of the Roses of the British Islands. *Annals of Scottish natural History* **13**:39-47.

CRÉPIN, F. 1896a. Révision des *Rosa* de l'herbier Babington. *Journal of Botany* **34**:178-182,212-216,266-270.

CRÉPIN, F. 1896b. *Rosae Americanae*. 1. *Botanical Gazette* **22**(1):1-34.

CRÉPIN, F. 1897. Le question de la priorité des noms spécifiques envisagée au point de vue du genre *Rosa*. *Bulletin de l'Herbier Boissier* **5**:129-163.

DAVIES, H. 1813. *Welsh Botanology*:49-53. London.

DENISE, M. 1903. *Rosa* L. In Coste, H. *Flore descriptive et illustrée de la France* **2**:47-57. Paris.

DÉSÉGLISE, A. 1864. *Rosa Bakeri* Déségl. mss. *Journal of Botany* **2**:267-269.

DÉSÉGLISE, A. 1864-5. Observations on the different methods proposed for the classification of the species of the genus *Rosa* L. *The Naturalist* **1**:273-276,292-298,308-313.

DÉSÉGLISE, A. 1865. Observations on Baker's "Review of the British Roses." *Journal of Botany* **3**:9-11.

DÉSÉGLISE, A. 1867. Revision of the section *Tomentosa* of the genus *Rosa*. *Journal of Botany* **5**:34-46,76-79.

DÉSÉGLISE, A. 1876. Catalogue raisonné ou énumeration méthodique des espèces du genre Rosier pour l'Europe, l'Asie et l'Afrique, specialement les Rosiers de la France et de l'Angleterre. *Bulletin de la Société royale de Botanique de Belgique* **15**(2):176-405,491-602.

DESPORTES, N.H.F. 1828. *Rosetum gallicum, ou énumeration méthodique des espèces et variétés du genre Rosier, indigènes en France ou cultivées dans les jardins, avec la synonymie française et latine.* Le Mans & Paris.

DESVAUX, A.N. 1809. Essai sur la géographie botanique du Haut-Poitou (Département de la Vienne). *Journal de Botanique* **2**:290-318.

DICKINSON, R. 1875. *Rosa spinosissima* var. *turbinata*. *Science Gossip* **9**:208.

DRUCE, G.C. 1878. *Rosa mollis* Sm. etc. in Northamptonshire. *Journal of Botany* **16**:25.

Du MORTIER, B. 1867. Monographie des Roses de la Flore Belge. *Bulletin de la Société royale de Botanique de Belgique* **6**:237-297.

EASLEA, W. 1896. The hybridization of Roses. *National Rose Society's prize essay.* Royal Horticultural Society, London.

EDWARDS, G. 1975. *Wild and old garden Roses.* London.

EMBLETON, R. 1868. *Rosa micrantha* new to Northumberland and Durham [Berwickshire]. *Berwickshire Naturalists' Field Club Transactions* **5**:408.

ERLANSON, E.W. 1928. Ten new American species and varieties of *Rosa*. *Rhodora* **30**:109-121.

ERLANSON, E.W. 1929. Cytological conditions and evidences for hybridity in North American wild Roses. *Botanical Gazette* **87**:443-506.

ERLANSON, E.W. 1931. Sterility in wild Roses and in some species hybrids. *Journal of Genetics* **16**:75-96.

ERLANSON, E.W. 1933. North American wild roses. *American Rose Annual* **17**:83-90.

ERLANSON, E.W. 1934. Experimental data for a revision of the North American wild Roses. *Botanical Gazette* **96**:197-259.

EXELL, A.W. 1930. Some overlooked names of plants. *Rosa parvula* Jacques. *Journal of Botany* **68**:300-301.

FALCONER, R.W. 1838. On the ancient history of the Rose. *Annals of natural History* **1**:228-229.

FAWCETT, W. 1914. *Rosa sinica*. *Journal of Botany* **52**:184.

FOSTER-MELLIAR, A. 1905. *The Book of the Rose*. Ed. 3. London.

FRASER, J. 1930. New records for Roses in Surrey. *Journal of Botany* **68**:154.

FRIES, E.M. 1814-23. *Novitiae Florae suecicae*. **1814**:1-40, **1817**:43-60, **1819**:63-80, **1823**:83-122. Lundae.

FRIES, E.M. & BRUNING, A.W. 1818. *Flora hallandica*: 84-87. Lund.

GANDOGER, M. 1892-93. *Monographia Rosarum Europae et Orientis*. **(1)**:1-338; **(2)**:1-486; **(3)**:1-418; **(4)**:1-601. Paris.

GAULT, S.M. & SYNGE, P.M. 1971. *The dictionary of Roses in colour*. London.

GORNALL, R.J. 1988. Specimens of data sheets from the British Rose survey. *B.S.B.I. News* **48**:14-15.

GORNALL, R.J. & PRIMAVESI, A.L. 1989. The British Rose survey of 1952-54. *Watsonia* **17**(3):356-359.

GRAHAM, G.G. & PRIMAVESI, A.L. 1990. Notes on some *Rosa* taxa recorded as occurring in the British Isles. *Watsonia* **18**:119-124.

GRENIER, J.C.M. 1865. *Flore de la Chaîne Jurassique*: 220-252. Paris.

GRENIER, J.C.M. 1876. Note on *Rosa glauca* Vill. *Bulletin de la Société royale de Botanique de Belgique* **15**(2):302-303.

GRENIER, J.C.M. & GODRON, D.A. 1848. *Flore de France*. Paris. **1**:551-561.

GUSTAFSSON, A. 1944. The constitution of the *Rosa canina* complex. *Hereditas* **30**:405-428, Fig.

GUSTAFSSON, A. & HAKANSSON, A. 1942. Meiosis in some *Rosa*-hybrids. *Botaniska Notiser* **1942**:331-343.

HARRISON, J.W.H. 1915. *Rosa eminens* Harrison – a new microgene of *Rosa mollissima* Willd. *The Vasculum* **1**:99-101.

HARRISON, J.W.H. 1916. The wild Roses of Durham. *The Naturalist* **1916**:9-13.

HARRISON, J.W.H. 1921. The genus *Rosa*, its hybridology and other genetical problems. *Transactions of the natural History Society of Northumberland, Durham and Newcastle-upon-Tyne* n.s. **5**(2):244-326.

HARRISON, J.W.H. 1929. On the variation of the Burnet Rose (*Rosa spinosissima*) in Northumberland & Durham. *The Vasculum* **15**:18-20.

HARRISON, J.W.H. 1930a. *Rosa spinosissima* var. *turbinata* Lind. and *R. tomentosa* Sm. *The Vasculum* **16**:162.

HARRISON, J.W.H. 1930b. A new species of Rose from Scotland and the North of England. *The Vasculum* **16**:144-146.

HARRISON, J.W.H. 1930c. The Roses of Winch's works. *The Vasculum* **16**:1-5.

HARRISON, J.W.H. 1930d. A revision of certain northern Rose groups. *The Naturalist* **1930**:161-167.

HARRISON, J.W.H. 1930e. New British Roses from Northumberland. *University of Durham philosophical Society Proceedings* **8**:161-167.

HARRISON, J.W.H. 1932. A new-old Rose, *Rosa mollis* var. *fallax* Harrison. *The Vasculum* **18**:23-24.

HARRISON, J.W.H. 1950. A new hybrid Rose found near Birtley. *The Vasculum* **35**(1):7.

HARRISON, J.W.H. 1954a. A new subspecies of Rose occurring in Durham. *The Vasculum* **39**:32-33.

HARRISON, J.W.H. 1954b. The wild Roses of Northumberland and Durham. *History and Transactions of the Consett & District Naturalists' Field Club* **1**:1-11.

HARRISON, J.W.H. 1955. Durham Wild Roses. *Proceedings of the botanical Society of the British Isles* **1**(3):369-371,373-374.

HARRISON, J.W.H. & BLACKBURN, K.B. 1927. The course of pollen production in certain Roses, with some deductions therefrom. *Memoirs of the horticultural Society of New York* **3**:23-32.

HARRISON, J.W.H. & BOLTON, E. 1938. The Rose flora of the Inner & Outer Hebrides and of other Scottish islands. *Transactions of the botanical Society of Edinburgh* **32**:424-431.

HERRING, P. 1925. Classification of *Rosa*. *Dansk botanisk Arkiv* **4**(9):1-24.

HERRMANN, J. 1762. *Dissertatio inauguralis botanico-medica de Rosa.* Argentorati.

HOBKIRK, C.P. 1867. Monsieur A. Déséglise's revision of the section *Tomentosae* of the genus *Rosa. The Naturalist* **3**:127-130,143-147.

HOFMANN, H. 1898. *Rosa Schlimperti* n.f., *R. canina* L. var. *dumalis* (Bechst.). *Allgemeine botanische Zeitschrift* **12**:192-193.

HOLE, S.R. 1872. *A book about Roses.* Ed. 4. Edinburgh & London.

HOGG, J. 1859. Notes on *Rosa rubella. Transactions of the Tyneside Naturalists' Field Club* **4**:185-187.

HOOKER, J. 1835. *British Flora* Ed. 3, **1**:223-243.

HURST, C.C. 1925. Chromosomes and characters in *Rosa* and their significance in the origin of species. In Harris, R.M. (ed.) *Experiments in Genetics*: 534-550, figs 169-175. Ed. 3. Cambridge.

HURST, C.C. 1928. Differential polyploidy in the genus *Rosa* L. *Zeitschrift für induktive Abstammungs und Vererbungslehre (Supplement)* **2**:866-906.

HURST, C.C. 1929. Genetics of the Rose. *The Rose Annual* **1929**:37-64. plates.

HURST, C.C. 1941. Notes on the origin and evolution of our garden Roses. *Journal of the Royal Horticultural Society* **66**:73-82, 242-250, 282-289.

HURST, C.C. & BREEZE, M.S.G. 1922. Notes on the origin of the Moss-Rose. *Journal of the Royal Horticultural Society* **47**:26-42.

HUTCHINSON, J. 1964. *The genera of flowering plants* **1**:174-216. Oxford.

JIČÍNSKÁ, D. 1976a. Autogamy in various species of the genus *Rosa. Preslia* **48**:225-229.

JIČÍNSKÁ, D. 1976b. Morphological features of F1 generation in Rose hybrids. 1. Hybrids of some species of the section *Caninae* with *Rosa rugosa. Folia geobotanica & phytotaxonomica* **11**:301-311.

KEAYS, E.E. 1935. *Old Roses.* New York.

KELLER, R. 1900-1905. *Rosa.* In Ascherson, P.F.A. & Graebner, K.O.P.P. *Synopsis der Mitteleuropäischen Flora* **6**(1):32-384. Leipzig.

KELLER, R. 1931. *Synopsis Rosarum spontanearum Europae mediae.* Zürich.

KELLER, R. & GAMS, H. 1923. *Rosa.* In Hegi, G. (ed.) *Illustrierte Flora von Mitteleuropa* **4**(2):976-1052. München.

KENT, D.H. 1992. *List of Vascular Plants of the British Isles.* London.

KLÁŠTERSKÝ, I. 1967. Les formes de *Rosa pimpinellifolia* L. *Bulletin de la Société royale de Botanique de Belgique* **37**:37-38.

KLÁŠTERSKÝ, I. 1968. *Rosa* L. In Tutin, T.G. et al. (eds.) *Flora Europaea* **2**:25-32. Cambridge.

KLÁŠTERSKÝ, I. 1969. Cytology and some chromosome numbers of Czechoslovak Roses, 1. *Folia geobotanica & phytotaxonomica* **4**:175-189.

KLÁŠTERSKÝ, I. 1971. New phenomena during meiosis in the genus *Rosa* L. *Hereditas* **67**(1): 55-63.

KLÁŠTERSKÝ, I. 1976. *Rosa arvensis* in der Tschechoslowakei. *Preslia* **48**:307-327.

KONČALOVÁ, M.N. 1971. Anthocyanidins from hips of *Rosa pimpinellifolia* L. *Preslia* **43**(3):198-201.

KONČALOVÁ, M.N. 1972. The frequency of species in a natural population of Roses from the Subsection *Eucaninae*. *Folia geobotanica & phytotaxonomica* **7**:423-424.

KONČALOVÁ, M.N. & KLÁŠTERSKÝ, I. 1978. Cytology and chromosome numbers of some Czechoslovak Roses, 3. *Folia geobotanica & phytotaxonomica* **13**:67-93.

KRÜSSMANN, G. 1981. *The complete Book of Roses*. Oregon.

LEES, A. 1919. The nightmare of names in the bed of Roses. *The Naturalist* **1919**:211-213.

LÉMAN, D.S. 1818. Extrait d'un mémoire de M. Léman. *Bulletin de Sciences par la Société Philomatique de Paris*:93-94.

LETT, H.W. 1907. Note on *Rosa hibernica*. *Journal of Botany* **45**:346-347.

LEWIS, W.H. 1958. A monograph of the genus *Rosa* in North America 2. *Southwestern Naturalist* **3**:145-174.

LEWIS, W.H. 1970. Species Roses in the United States: their relationship to modern Roses. *American Rose Annual* **55**:78-85.

LEY, A. 1907. British roses of the *Mollis-Tomentosa* group. *Journal of Botany* **45**:200-210.

LEY, A. 1908a. The *Villosae* section of the genus *Rosa*. *Journal of Botany* **46**:328-329.

LEY, A. 1908b. *Rosae villosae* in the 'London Catalogue'. *Journal of Botany* **46**:394.

LEY, A. 1908c. *Rosa pomifera* J. Herrm. as British. *Journal of Botany* **46**:58.

LEY, A. & WOLLEY-DOD, A.H. 1909. The collection and identification of Roses. *Journal of Botany* **47**:247-255.

LINDLEY, J. 1820. *Rosarum Monographia: or, a botanical History of Roses.* London.

LINDSTROM, A.A. 1926. *Rosa* L. In Lindman, C.A.M. *Svensk Fanerogamflora*: 361-373. Stockholm.

LINNAEUS, C. 1762. *Species Plantarum.* Ed.2:703-705. Holmiae.

McFARLAND, J.H. 1936. *Roses of the World in Colour.* Boston.

MACVICAR, S.M. 1895. *Rosa mollis* Sm. var. *glabrata* Fries. *Journal of Botany* **33**:344-345.

MALMGREN, J. 1986. Släktet *Rosa* i Sverige. *Svensk botanisk Tidskrift* **80**:209-227.

MARSHALL, E.S. 1895. Two additions to the list of British Roses. *Journal of Botany* **33**:43-45.

MATTHEWS, J.R. 1910. On some British hybrid Roses. *Transactions of the botanical Society of Edinburgh* **24**(3):135-142, pl.12.

MATTHEWS, J.R. 1920. Hybridism and classification in the genus *Rosa. New Phytologist* **19**(7&8):153-171.

MATTHEWS, J.R. 1926a. Notes on Fife and Kinross Roses. *Transactions of the botanical Society of Edinburgh* **29**(3):219-225.

MATTHEWS, J.R. 1926b. *The Roses of Britain.* By Lt.-Col. A.H. Wolley-Dod. [Review.] *Journal of Botany* **64**:354-356.

MATTHEWS, J.R. 1934. *Rosa Perthensis* Rouy and its history as a British plant. *Journal of Botany* **72**:167-171.

MATTHEWS, J.R. 1976. What is × *Rosa Perthensis* Rouy? *Botanical Society of Edinburgh News* **18**:15-16.

MELVILLE, R. 1952. Field Meetings, 1951. Halling Down, Kent. *Year Book of the Botanical Society of the British Isles* **1952**:44-46.

MELVILLE, R. 1960. A metrical study of leaf-shape in hybrids. *Kew Bulletin* **14**:88-102,161-177.

MELVILLE, R. 1967. The problem of classification in the genus *Rosa. Bulletin de la Société royale de Botanique de Belgique* **37**:39-44.

MELVILLE, R. 1975. *Rosa* L. In Stace, C.A. (ed.) *Hybridization and the Flora of the British Isles*:212-227. London.

MELVILLE, R. & PYKE, M. 1942. Vitamin C in Rose hips. *Biochemical Journal* **36**:336-339.

MÉRAT, F.V. 1812. *Nouvelle Flore des Environs de Paris*:189-193. Paris.

MOORE, D.M. 1987. *Rosa* L. In Clapham, A.R. et al. *Flora of the British Isles* Ed.3:225-230. Cambridge.

MOYLE ROGERS, W. 1889. *Rosa stylosa* var. *pseudorusticana* Crép. *Journal of Botany* **27**:23-24.

NICHOLSON, G. 1887. *Rosa ripartii* Déséglise in Britain. *Journal of Botany* **25**:111.

NIESCHALK, A. & NIESCHALK, C. 1975. Beiträge zur Kenntnis der Rosenflora Nordhessens, 1. Der Formenkreise um *Rosa elliptica* Tausch. *Philippia* **2**:299-316.

NIESCHALK, A. & NIESCHALK, C. 1978. Beiträge zur Kenntnis der Rosenflora Nordhessens, 2. Der Formenkreise um *Rosa agrestis* Savi. *Philippia* **3**:389-407.

NIESCHALK, A. & NIESCHALK, C. 1980. Beiträge zur Kenntnis der Rosenflora Nordhessens, 3. Der Formenkreise um *Rosa micrantha* Borrer ex Sm. *Philippia* **4**:213-233

NIESCHALK, A. & NIESCHALK, C. 1981. Beiträge zur Kenntnis der Rosenflora Nordhessens, 4. Der Formenkreise um *Rosa rubiginosa* L. *Philippia* **4**:388-413.

NIESCHALK, A. & NIESCHALK, C. 1986. Beiträge zur Kenntnis der Rosenflora Nordhessens, 5. Der Formenkreise um *Rosa tomentosa* Sm., *R. scabriuscula* Sm., *R. villosa* L., *R. sherardii* Davies. *Philippia* **5**:318-345.

NILSSON, Ö. 1967. Drawings of Scandinavian plants. *Rosa* L. *Botaniska Notiser* **120**:1-8,137-143,249-254,393-408.

O'MAHONY, T. 1977. Current taxonomic and distributional research on the genus *Rosa* in Ireland. *Irish Biogeographical Society Bulletin* **1**:41-44.

PARMENTIER, P.A. 1897. Recherches anatomiques et taxonomiques sur les Rosiers. *Annales des Sciences naturelles* sér.8. **6**:1-175.

PAYNE, C.H. 1922. The history of the Moss-Rose, a critique. *Gardeners' Chronicle & agricultural Gazette* **1922**:48,69-70,84,93,108,124,135.

PRAEGER, R.Ll. 1928. A new hybrid Rose. *Journal of Botany* **66**:87-88.

PRAEGER, R.Ll. 1934. The standing of certain plants in Ireland. *Rosa stylosa* Desv. *Journal of Botany* **72**:69-70.

PRIMAVESI, A.L. 1992. The *Rosa* herbarium of A.H. Wolley-Dod. *Watsonia* **19**:137-139.

PRIMAVESI, A.L. & EVANS, P.A. 1988. *Flora of Leicestershire*. Leicester.

RAU, A. 1816. *Enumeratio Rosarum circa Wirceburgum et pagos adjacentes sponte crescentium*. Norimbergiae.

REHDER, A. 1947. *Manual of the cultivated trees and shrubs, hardy in North America*. Ed. 2. New York.

REICHERT, H. 1986. Kritische Anmerkungen zur Beschreibung und Verschlüsselung der engeren *Rosa canina* Gruppe in der *Flora Europaea*. *Gottinger Floristische Rundbriefe* **19**:66-70.

RENDLE, H.H. 1928. Short notes: American Roses – Mrs E.W. Erlanson. *Journal of Botany* **66**:274-275.

ROBERTS, A.V. 1977. Relationship between species in the genus *Rosa*, Section *Pimpinellifoliae*. *Botanical Journal of the Linnean Society* **74**:309-328.

ROUY, G. 1899. *Rosa* L. In Rouy, G. & Foucaud, J. *Flore de la France* **6**:236-241. Asnières, Paris & Rochefort.

ROWLEY, G.D. 1956. Roses at Bayfordbury. *The Rose Annual* **1956**:11-17.

ROWLEY, G.D. 1959. Ancestral China Roses. *Journal of the Royal Horticultural Society* **84**(6):270-274.

ROWLEY, G.D. 1961. The Scotch Rose and its garden descendents. *Journal of the Royal Horticultural Society* **86**:544-437.

ROWLEY, G.D. 1960. Aneuploidy in the genus *Rosa*. *Journal of Genetics* **57**:253-268.

ROWLEY, G.D. 1967. Chromosome studies and evolution in the genus *Rosa*. *Bulletin de la Société royale de Botanique de Belgique* **37**:45-52.

ROWLEY, G.D. 1976. Typification of the genus *Rosa*. *Taxon* **25**(1): 169-210.

RUTHERFORD, A. 1990. The herbaceous border – 2. *Rosa rugosa* and its hybrid. *BSBI News* **55**:24-27.

SABINE, J. 1837. Notice respecting a native British Rose, first described in Ray's Synopsis as discovered by James Sherard. *Transactions of the Linnean Society of London* **17**(4):539-541.

SAVI, G. 1798. *Flora pisana*. **1**:471-478. Pisa.

SCHENK, E. 1955 & 1957. Bestimmungsflora der Deutschen Wildrosen. *Mitteilungen der floristisch-soziologischen Arbeitsgemeinschaft* **5 & 6/7**.

SCHEUTZ, N.J. 1888. De duabus Rosis britannicis. *Journal of Botany* **26**:67-68.

SCHULTZ, F.W. 1842-1855. *Archives de la Flore de France et d'Allemagne*. Bitche, Haguenau & Deux-Ponts.

SCHWERTSCHLAGER, J. 1910. *Die Rosen des südlichen und mittleren Frankenjura*. München.

SERINGE, N.C. 1825. *Rosa* L. In De Candolle, A.P. *Prodromus systematis naturalis Regni vegetabilis* **2**:597-625. Argentorati & Londini.

SIM, J. 1864-65. *Rosa alpina* L. in Britain. *The Naturalist* **1**:184-186.

STACE, C.A. 1991. *New flora of the British Isles*:426-437. Cambridge.

STEARN, W.T. 1966. *Botanical Latin*. London.

STOCK, K.L. 1984. *Rose Books. A Bibliography of books and important articles in journals on the genus Rosa, in English, French, German, and Latin 1550-1975*. Milton Keynes.

SYLVESTER-BRADLEY, P.C. 1952. *British Rose survey, instructions, charts, table, etc*. Circulated privately.

SYLVESTER-BRADLEY, P.C. 1953. The British Rose survey. *Year-book of the Botanical Society of the British Isles* **1953**:66.

SYLVESTER-BRADLEY, P.C. 1954-55. The taxonomic implications of the British Rose survey. *Proceedings of the botanical Society of the British Isles* **1**:254-255.

TÄCKHOLM, G. 1920. On the cytology of the genus *Rosa*. A preliminary note. *Svensk botanisk Tidskrift* **14**:300-311, 3 figs.

TÄCKHOLM, G. 1922. Zytologische Studien uber die Gattung *Rosa*. *Acta Horti bergiana* **7**(3):97-381.

THOMAS, G.S. 1971. *The old shrub Roses*. London.

THOMAS, G.S. 1980. *Rosa* L. In Bean, W.J. *Trees and Shrubs hardy in the British Isles*. Ed.8, **4**:36-205. London.

THUILLIER, J.L. 1790. *Flore des Environs de Paris*:140-141. Paris.

TRAAEN, C. 1911. Scandinavian Roses. *Journal of Botany* **49**:298-300.

TRAAEN, C. 1913. *Rosa Afzeliana* Fries. *Journal of Botany* **51**:127-129.

TRAAEN, C. 1920. Sveriges *Rosae* by S. Almquist, Stockholm, 1919. [Review.] *Journal of Botany* **58**:115-117.

TRATTINNICK, L. 1823-24. *Rosacearum Monographia*. Vindobonae.

VAUGHAN, I.M. 1966. *Rosa micrantha* Borrer ex Sm., *R. tomentosa* Sm., *R. afzeliana* Fr. & *R. sherardii* Davies. *Proceedings of the botanical Society of the British Isles* **6**:383.

VAUGHAN, I.M. 1985. Notes on the identification of species in the genus *Rosa* in Carmarthenshire. *Llanelli Naturalists' Newsletter*, June 1985.

VĚTVIČKA, V. 1972. Diagnostic and taxonomic importance of sepals in the genus *Rosa* L., 2. *Cas. Slezsk. Muz. c., Dendrol.* **2**:119-128.

VICIOSO, C. 1964. *Estudios sobre el genero 'Rosa' en Espana.* Ed.2. Madrid.

WARBURG, E.F. 1962. *Rosa* L. In Clapham, A.R. et al. *Flora of the British Isles.* Ed.2,405-413. Cambridge.

WEBB, D.A. 1951. Hybrid plants in Ireland. *Irish Naturalists' Journal* **10**:201-204.

WILLMOTT, E.A. 1910-1914. *The genus* Rosa. 2 vols. London.

WILMOTT, A.J. 1950. A new method for the identification and study of critical groups. *Proceedings of the Linnean Society of London* **162**:83-98.

WINCH, N.J. 1816. Some indigenous Roses. *Monthly Magazine* **1816**:291-293.

WOLLEY-DOD, A.H. 1908a. *Rosa obovata* Ley. *Journal of Botany* **46**:364.

WOLLEY-DOD, A.H. 1908b. The subsection *Eu-caninae* of the genus *Rosa*. *Journal of Botany* **46**(Supplement):1-110.

WOLLEY-DOD, A.H. 1910a. The British roses (excluding *Eu-canina.*) *Journal of Botany* **48**(Supplement):1-141.

WOLLEY-DOD, A.H. 1910b. Note on *Rosa*. *Journal of Botany* **48**:187-188.

WOLLEY-DOD, A.H. 1911. A list of British Roses. *Journal of Botany* **49**(Supplement):1-67.

WOLLEY-DOD, A.H. 1920a. On collecting Roses. *Journal of Botany* **58**:23-24.

WOLLEY-DOD, A.H. 1920b. A revised arrangement of British Roses. *Journal of Botany* **58**(Supplement):1-20.

WOLLEY-DOD, A.H. 1924a. Notes on collecting Roses. *Journal of Botany* **62**:52-53.

WOLLEY-DOD, A.H. 1924b. Some new British Roses. *Journal of Botany* **62**:202-209.

WOLLEY-DOD, A.H. 1924c. *The Roses of Britain.* London.

WOLLEY-DOD, A.H. 1925. Corrected names of Roses distributed in the past through the B.E.C. of the British Isles. *Report of the botanical Society and Exchange Club of the British Isles* **7**:667-684.

WOLLEY-DOD, A.H. 1928. *Rosa hibernica* Templeton. *Journal of Botany* **66**:361-362.

WOLLEY-DOD, A.H. 1929. On some varieties of *Rosa tomentosa (Scabriusculae). Journal of Botany* **67**:38-42,87.

WOLLEY-DOD, A.H. 1930a. *Rosa scabriuscula* Smith. *Journal of Botany* **68**:185-187.

WOLLEY-DOD, A.H. 1930b. A revision of certain northern Rose groups. *Journal of Botany* **68**:248-249.

WOLLEY-DOD, A.H. 1930-31. A revision of the British Roses. *Journal of Botany* **68**(Supplement), **69**(Supplement).

WOLLEY-DOD, A.H. 1931. *Rosa canina* f. *Wolley-Dodii* Sudre, *Rosa* × *Rogersii* W.-Dod, *Rosa tomentella* Lem. var. *decipiens. Report of the botanical Society and Exchange Club of the British Isles* **9**:827.

WOLLEY-DOD, A.H. 1936. Some Rose notes. *Report of the botanical Society and Exchange Club of the British Isles* **11**:68-81.

WOODS, J. 1818. A synopsis of the British species of *Rosa. Transactions of the Linnean Society of London* **12**:159-234.

WYLIE, A.P. 1954a. Variation in the genetical structure of the *Caninae* roses. *Proceedings of the botanical Society of the British Isles* **1**:103.

WYLIE, A.P. 1954b. The status of *Rosa wilsoni. Proceedings of the botanical Society of the British Isles* **1**:265.

WYLIE, A.P. 1954-55c. The history of the garden Rose. *Journal of the Royal horticultural Society* **79**:555-571, **80**:8-24,77-87.

ZIELIŃSKI, J. 1988. Studies in the genus *Rosa* L. – systematics of section *Caninae* DC. em. Christ. *Arboretum Kórnickie* **30**:3-109.

ZIELIŃSKI, J. 1987. *Rosa* L.– Roza. *Flora Polski*. Warsaw.

16. Index

The numerals refer to taxa, not to pages

Burnet Rose	5
Columnar-styled Dog-rose	11
Dog-rose	12
Columnar-styled	11
Glaucous Northern	13b
Hairy Northern	13a
Long-styled	11
Northern	13
Round-leaved	14
Downy-rose	
Harsh	15
Sherard's	16
Soft	17
Dutch Rose	7
Field-rose	4
Glaucous Northern Dog-rose	13b
Hairy Northern Dog-rose	13a
Harsh Downy-rose	15
Japanese Rose	6
Many-flowered Rose	1
Memorial Rose	3
Northern Dog-rose	13
Prairie Rose	2
Red-leaved Rose	8
Red Rose of Lancaster	10
Round-leaved Dog-rose	14
Sherard's Downy-rose	16
Short-styled Field-rose	11
Small-flowered Sweet-briar	19
Small-leaved Sweet-briar	20
Soft Downy-rose	17
Sweet-briar	18
Small-flowered	19
Small-leaved	20
Virginian Rose	9
White Rose of York	4 × 10
Rosa	
× *aberrans* W.-Dod	= 12 × 15, 15 × 12
afzeliana Fries	= 13b

agrestis Savi	Map 32, **20**
agrestis × *canina*	**20** × **12**
agrestis × *micrantha*	**20** × **19**
agrestis × *sherardii*	**20** × **16**
agrestis × *stylosa*	**20** × **11**
× *alba* L.	**4** × **10**
× *andegavensis* Bast.	**11** × **12, 12** × **11**
arvensis Huds.	Map 2, **4**
arvensis × *canina*	Map 4, **4** × **12**
arvensis × *gallica*	**4** × **10**
arvensis × *micrantha*	**4** × **19**
arvensis × *rubiginosa*	**4** × **18**
arvensis × *sherardii*	**4** × **16**
arvensis × *tomentosa*	**4** × **15**
arvensis × *stylosa*	Map 3, **4** × **11**
× *avrayensis* Rouy	**15** × **18, 18** × **15**
× *belnensis* Ozan	**20** × **12**
× *belnensis* auct., non Ozan	= **11** × **20, 20** × **11**
× *bibracteoides* W.-Dod	= **4** × **11, 11** × **4**
× *bigeneris* Duffort ex Rouy	**19** × **18**
× *bishopii* W.-Dod	**19** × **20, 20** × **19**
× *burdonii* W.-Dod	= **16** × **18, 18** × **16**
caesia Sm.	**13**
subsp. *caesia*	Map 22, **13a**
subsp. *glauca* (Nyman) G.G. Graham & Primavesi	Map 23, **13b**
caesia × *arvensis*	**13** × **4**
caesia × *canina*	Maps 17 & 18, **13** × **12**
caesia × *micrantha*	**13** × **19**
caesia × *mollis*	**13** × **17**
caesia × *pimpinellifolia*	**13** × **5**
caesia × *rubiginosa*	**13** × **18**
caesia × *sherardii*	Map 24, **13** × **16**
caesia × *tomentosa*	**13** × **15**
canina L.	Maps 13-16, **12**
canina × *arvensis*	Map 4, **12** × **4**
canina × *caesia*	Maps 17 & 18, **12** × **13**
canina × *micrantha*	**12** × **19**
canina × *mollis*	**12** × **17**
canina × *obtusifolia*	Map 19, **12** × **14**
canina × *pimpinellifolia*	Map 6, **12** × **5**
canina × *rubiginosa*	**12** × **18**
canina × *sherardii*	Map 21, **12** × **16**
canina × *stylosa*	Map 12, **12** × **11**

canina × tomentosa	12 × 15
× cantiana (W.-Dod) W.-Dod	5 × 18, 18 × 5
× collina Jacq., non J. Woods	= 4 × 10
× concinnoides W.-Dod	= 12 × 14, 14 × 12
× consanguinea Gren.	4 × 18, 18 × 4
coriifolia Fries	= 13a
× coronata Crépin ex Reuter	= 5 × 15, 15 × 5
corymbifera Borkh.	= 12
× curvispina W.-Dod	= 12 × 15, 15 × 12
× deseglisei Boreau	= 4 × 12, 12 × 4
dumalis auct., non Bechst.	= 13b, 12
× dumalis Bechst.	12 × 13, 13 × 12
dumetorum auct., non Thuill.	= 12
× dumetorum Thuill.	12 × 14, 14 × 12
gallica L.	10
glauca Pourret	8
glauca Villars ex Lois., non Pourret	= 13b
× glaucoides W.-Dod	13 × 17, 17 × 13
× hibernica Templeton	5 × 12, 12 × 5
'Hollandica'	7
× inelegans W.-Dod	4 × 19, 19 × 4
× involuta Sm.	5 × 16, 16 × 5
× kosinsciana Besser	= 4 × 12, 12 × 4
× latebrosa Besser	= 12 × 18, 18 × 12
× latens W.-Dod	= 12 × 18, 18 × 12
× longicolla Ravaud ex Rouy	13 × 19
luciae Franchet & Rochebr.	3
× margerisonii (W.-Dod) W.-Dod	5 × 13, 13 × 5
micrantha Borrer ex Sm.	Map 31, 19
micrantha × agrestis	19 × 20
micrantha × arvensis	19 × 4
micrantha × canina	19 × 12
micrantha × obtusifolia	19 × 14
micrantha × rubiginosa	19 × 18
micrantha × stylosa	19 × 11
× molletorum H.-Harr.	12 × 17, 17 × 12
× molliformis W.-Dod	17 × 18, 18 × 17
mollis Sm.	Map 29, 17
mollis × caesia	17 × 13
mollis × canina	17 × 12
mollis × pimpinellifolia	Map 8, 17 × 5
mollis × rubiginosa	17 × 18
mollis × sherardii	Map 28, 17 × 16

× *moorei* (Baker) W.-Dod	= 5 × 18, 18 × 5
multiflora Thunb. ex Murray	1
× *nitidula* Besser	12 × 18, 18 × 12
× *obovata* (Baker) Ley, non Raf.	= 13 × 18, 18 × 13
obtusifolia Desv.	Map 25, 14
obtusifolia × *arvensis*	14 × 4
obtusifolia × *caesia*	14 × 13
obtusifolia × *canina*	Map 19, 14 × 12
obtusifolia × *rubiginosa*	14 × 18
obtusifolia × *stylosa*	14 × 11
obtusifolia × *tomentosa*	Map 20, 14 × 15
× *perthensis* Rouy	= 5 × 16
pimpinellifolia L.	Map 5, 5
pimpinellifolia × *caesia*	5 × 13
pimpinellifolia × *canina*	Map 6, 5 × 12
pimpinellifolia × *mollis*	Map 8, 5 × 17
pimpinellifolia × *rubiginosa*	Map 9, 5 × 18
pimpinellifolia × *sherardii*	Map 7, 5 × 16
pimpinellifolia × *tomentosa*	5 × 15
× *praegeri* W.-Dod	6 × 12
× *pseudorusticana* Crépin ex Rogers	4 × 11, 11 × 4
× *rogersii* W.-Dod	13 × 15, 15 × 13
× *rothschildii* Druce	Map 21, 12 × 16, 16 × 12
× *rouyana* Duffort ex Rouy	15 × 4
rubiginosa L.	Map 30, 18
rubiginosa × *arvensis*	18 × 4
rubiginosa × *canina*	18 × 12
rubiginosa × *mollis*	18 × 17
rubiginosa × *pimpinellifolia*	Map 9, 18 × 5
rubiginosa × *sherardii*	18 × 16
rubiginosa × *stylosa*	18 × 11
rubiginosa × *tomentosa*	18 × 15
rubrifolia Villars, nom. illegit.	= 8
× *rufescens* W.-Dod	= 11 × 12, 12 × 11
rugosa Thunb. ex Murray	Map 10, 6
rugosa × *canina*	6 × 12
× *sabinii* Woods	5 × 17, 17 × 5
scabriuscula auct., non Sm.	= 15
× *scabriuscula* Sm.	12 × 15, 15 × 12
setigera Michaux	2
× *setonensis* W.-Dod	= 5 × 13, 13 × 5
sherardii Davies	Map 27, 16
sherardii × *agrestis*	16 × 20

sherardii × *arvensis*	**16** × **4**
sherardii × *caesia*	Map 24, **16** × **13**
sherardii × *canina*	Map 21, **16** × **12**
sherardii × *micrantha*	**16** × **19**
sherardii × *mollis*	Map 28, **16** × **17**
sherardii × *pimpinellifolia*	Map 7, **16** × **5**
sherardii × *rubiginosa*	**16** × **18**
sherardii × *tomentosa*	**16** × **15**
× *shoolbredii* W.-Dod	**16** × **17**, **17** × **16**
spinosissima L.	= **5**
squarrosa auct., ?(Rau) Boreau	= **12**
stylosa Desv.	Map 11, **11**
stylosa × *agrestis*	**11** × **20**
stylosa × *arvensis*	Map 3, **11** × **4**
stylosa × *caesia*	**11** × **13**
stylosa × *canina*	Map 12, **11** × **12**
× *subcanina* (Christ) Dalla Torre & Sarnth.	= **12** × **13**, **13** × **12**
× *subcollina* (Christ) Dalla Torre & Sarnth.	= **12** × **13**, **13** × **12**
× *suberecta* (Woods) Ley	**16** × **18**, **18** × **16**
× *suberectiformis* W.-Dod	**15** × **16**, **16** × **15**
× *subobtusifolia* W.-Dod	= **12** × **4**, **4** × **12**
× *timbalii* Rouy	= **15** × **18**, **18** × **15**
× *toddiae* W.-Dod	**12** × **19**, **19** × **12**
× *tomentelliformis* W.-Dod	**14** × **18**
tomentosa Sm.	Map 26, **15**
tomentosa × *agrestis*	**15** × **20**
tomentosa × *caesia*	**15** × **13**
tomentosa × *canina*	**15** × **12**
tomentosa × *micrantha*	**15** × **19**
tomentosa × *mollis*	**15** × **17**
tomentosa × *obtusifolia*	Map 20, **15** × **14**
tomentosa × *pimpinellifolia*	**15** × **5**
tomentosa × *rubiginosa*	**15** × **18**
tomentosa × *sherardii*	**15** × **16**
× *verticillacantha* Mérat	**4** × **12**, **12** × **4**
villosa auct., non L.	**17**
virginiana Herrm.	**9**
vosagiaca N. Desp.	= **13b**
× *wheldonii* W.-Dod	= **4** × **12**, **12** × **4**
wichuraiana Crépin	= **3**

BSBI Handbooks

Each handbook deals in depth with one of the more difficult groups of British and Irish plants.

No. 1 SEDGES OF THE BRITISH ISLES
A.C. Jermy, A.O. Chater ad R.W. David. 1982. 268 pages, with a line drawing and distribution map for every British species. Paperback. ISBN 0 901158 05 4

No. 2 UMBELLIFERS OF THE BRITISH ISLES
T.G. Tutin. 1980. 197 pages, fully illustrated with line drawings for each species. Paperback. ISBN 0 901158 02 X. Out of print. 2nd Edn in preparation. Orders being recorded.

No. 3 DOCKS AND KNOTWEEDS OF THE BRITISH ISLES
J.E. Lousley and D.H. Kent. 1981. 205 pages, with many line drawings of British native and alien taxa. Paperback. ISBN 0 901158 04 6. Out of print. 2nd Edn in preparation. Orders being recorded.

No. 4 WILLOWS AND POPLARS OF GREAT BRITAIN AND IRELAND
R.D. Meikle. 1984. 198 pages with 63 line drawings of all the British and Irish species, subspecies, varieties and hybrids. Paperback. ISBN 0 901158 07 0

No. 5 CHAROPHYTES OF GREAT BRITAIN AND IRELAND
J.A. Moore. 1986. 144 pages with line drawings and maps. Paperback. ISBN 0901158 16 X

No. 6 CRUCIFERS OF GREAT BRITAIN AND IRELAND
T. C. G. Rich. 1991. 336 pages with 135 line drawings covering 136 species. 60 distribution maps. Paperback. ISBN 0 901158 20 8

Available from the official agents for B.S.B.I. Publications:
F & M Perring
24 Glapthorn Road
Oundle
Peterborough PE8 4JQ
England